PRACTICAL TALKS BY AN ASTRONOMER

The Moon. First Quarter.
Photographed by Loewy and Puiseux, February 13, 1894.

PRACTICAL TALKS BY AN ASTRONOMER

BY

HAROLD JACOBY
ADJUNCT PROFESSOR OF ASTRONOMY IN
COLUMBIA UNIVERSITY

ILLUSTRATED

PREFACE

The present volume has not been designed as a systematic treatise on astronomy. There are many excellent books of that kind, suitable for serious students as well as the general reader; but they are necessarily somewhat dry and unattractive, because they must aim at completeness. Completeness means detail, and detail means dryness.

But the science of astronomy contains subjects that admit of detached treatment; and as many of these are precisely the ones of greatest general interest, it has seemed well to select several, and describe them in language free from technicalities. It is hoped that the book will thus prove useful to persons who do not wish to give the time required for a study of astronomy as a whole, but who may take pleasure in devoting a half-hour now and then to a detached essay on some special topic.

Preparation of the book in this form has made it suitable for prior publication in periodicals; and the several essays have in fact all been printed before. But the intention of collecting them into a book was kept in mind from the first; and while no attempt has been made at consecutiveness, it is hoped that nothing of merely ephemeral value has been included.

CONTENTS

Navigation at Sea
The Pleiades
The Pole-Star
Nebulæ
Temporary Stars
Galileo
The Planet of 1898
How to Make a Sun-Dial
Photography in Astronomy
Time Standards of the World
Motions of the Earth's Pole
Saturn's Rings
The Heliometer
Occultations
Mounting Great Telescopes
The Astronomer's Pole
The Moon Hoax
The Sun's Destination

ILLUSTRATIONS

THE MOON. FIRST QUARTER
Photographed by Loewy and Puiseux, February 13, 1894.

SPIRAL NEBULA IN CONSTELLATION LEO
Photographed by Keeler, February 24, 1900.

NEBULA IN ANDROMEDA
Photographed by Barnard, November 21, 1892.

THE "DUMB-BELL" NEBULA
Photographed by Keeler, July 31, 1899.
STAR-FIELD IN CONSTELLATION MONOCEROS
Photographed by Barnard, February 1, 1894.

SOLAR CORONA. TOTAL ECLIPSE
Photographed by Campbell, January 22, 1898; Jeur, India.

FORTY-INCH TELESCOPE, YERKES OBSERVATORY
YERKES OBSERVATORY, UNIVERSITY OF CHICAGO

NAVIGATION AT SEA

A short time ago the writer had occasion to rummage among the archives of the Royal Astronomical Society in London, to consult, if possible, the original manuscripts left by one Stephen Groombridge, an English astronomer of the good old days, who died in 1832. It was known that they had been filed away about a generation ago, by the late Sir George Airy, who was Astronomer Royal of England between the years 1835 and 1881. After a long search, a large and dusty box was found and opened. It was filled with documents, of which the topmost was in Sir George's own handwriting, and began substantially as follows:

"List of articles within this box.

"No. 1, This list,
"No. 2, etc., etc."

Astronomical precision can no further go: he had listed even the list itself. Truly, Airy was rightly styled "prince of precisians." A worthy Astronomer Royal was he, to act under the royal warrant of Charles II., who established the office in 1675. Down to this present day that warrant still makes it the duty of His Majesty's Astronomer "to apply himself with the most exact care and diligence to the rectifying of the tables of the motions of the heavens and the places of the fixed stars, in order to find out the so much desired longitude at sea, for the perfecting the art of navigation."

The "so much desired longitude at sea" is, indeed, a vastly important thing to a maritime nation like England. And only in comparatively recent years has it become possible and easy for vessels to be navigated with safety and convenience upon long voyages. The writer was well acquainted with an old sea-captain of New York, who had commanded one of the earliest transatlantic steamers, and who died only a few years ago. He had a goodly store of ocean yarn, fit and ready for the spinning, if he could but find someone who, like himself, had known and loved the ocean. In his early sea-going days, only the wealthiest of captains owned chronometers. This instrument is now considered indispensable in navigation, but at that time it was a new invention, very rare and costly. Upon a certain voyage from England to Rio Janeiro, in South America, the old captain could remember the following odd method of navigation: The ship was steered by compass to the southward and westward, more or less, until the skipper's antique quadrant showed that they had about reached the latitude of Rio. Then they swung her on a course due west by compass, and away she went for Rio, relying on the lookout man forward to keep the ship from running ashore. For after a certain lapse of time, being ignorant of the longitude, they could not know whether they would "raise" the land within an hour or in six weeks. We are glad of an opportunity to put this story on record, for the time is not far distant when there will be no man left among the living who can remember how ships were taken across the seas in the good old days before chronometers.

Anyone who has ever been a passenger on a great transatlantic liner of to-day knows what an important, imposing personage is the brass-bound skipper. A very different creature is he on the deck of his ship from the modest seafaring man we meet on land, clad for the time being in his shore-going togs. But the captain's dignity is not all brass buttons and gold braid. He has behind him the powerful support of a deep, delightful mystery. He it is who "takes the sun" at noon, and finds out the ship's path at sea. And in truth, regarded merely as a scientific experiment, the guiding of a vessel across the unmarked trackless ocean has few equals within the whole range of human knowledge. It is our purpose here to explain quite briefly the manner in which this seeming impossibility is accomplished. We shall not be able to go sufficiently into details to enable him who reads to run and navigate a magnificent steamer. But we hope to diminish somewhat that small part of the captain's vast dignity which depends upon his mysterious operations with the sextant.

To begin, then, with the sextant itself. It is nothing but an instrument with which we can measure how high up the sun is in the sky. Now, everyone knows that the sun slowly climbs the sky in the morning, reaches its greatest height at noon, and then slowly sinks again in the afternoon. The captain simply begins to watch the sun through the sextant shortly before noon, and keeps at it until he discovers that the sun is just beginning to descend. That is the instant of noon on the ship. The captain quickly glances at the chronometer, or calls out "noon" to an officer who is near that instrument. And so the error of the chronometer becomes known then and there without any further astronomical calculations whatever. Navigators can also find the chronometer error by sextant observations when the sun is a long way from noon. The methods of doing this are somewhat less simple than for the noon observation; they belong to the details of navigation, into which we cannot enter here.

Incidentally, the captain also notes with the sextant how high the sun was in the sky at the noon observation. He has in his mysterious "chart-room" some printed astronomical tables, which tell him in what terrestrial latitude the sun will have precisely that height on that particular day of the year. Thus the terrestrial latitude becomes known easily enough, and if only the captain could get his longitude too, he would know just where his ship was that day at noon.

We have seen that the sextant observations furnish the error of the chronometer according to ship's time. In other words, the captain is in possession of the correct local time in the place where the ship actually is. Now, if he also had the correct time at that moment of some well-known place on shore, he would know the difference in time between that place on shore and the ship. But every traveller by land or sea is aware that there are always differences of time between different places on the earth. If a watch be right on leaving New York, for instance, it will be much too fast on arriving at Chicago or San Francisco; the farther you go the larger becomes the error of your watch. In fact, if you could find out how much your watch had gone into error, you would in a sense know how far east or west you had travelled.

Now the captain's chronometer is set to correct "Greenwich time" on shore before the ship leaves port. His observations having then told him how much this is wrong on that particular day, and in that particular spot where the ship is, he knows at once just how far he has travelled east or west from Greenwich.

In other words, he knows his "longitude from Greenwich," for longitude is nothing more than distance from Greenwich in an east-and-west direction, just as latitude is only distance from the equator measured in a north-and-south direction. Greenwich observatory is usually selected as the beginning of things for measuring longitudes, because it is almost the oldest of existing astronomical establishments, and belongs to the most prominent maritime nation, England.

One of the most interesting bits of astronomical history was enacted in connection with this matter of longitude. From what has been said, it is clear that the ship's longitude will be obtained correctly only if the chronometer has kept exact time since the departure of the ship from port. Even a very small error of the chronometer will throw out the longitude a good many miles, and we can understand readily that it must be difficult in the extreme to construct a mechanical contrivance capable of keeping exact time when subjected to the rolling and pitching of a vessel at sea.

It was as recently as the year 1736 that the first instrument capable of keeping anything like accurate time at sea was successfully completed. It was the work of an English watchmaker named John Harrison, and is one of the few great improvements in matters scientific which the world owes to a desire for winning a money prize. It appears that in 1714 a committee was appointed by the House of Commons, with no less a person than Sir Isaac Newton himself as one of its members, to consider the desirability of offering governmental encouragement for the invention of some means of finding the longitude at sea. Finally, the British Government offered a reward of $50,000 for an instrument which would find the longitude within sixty miles; $75,000, if within forty miles, and $100,000, if within thirty miles. Harrison's chronometer was finished in 1736, but he did not receive the final payment of his prize until 1764.

We shall not enter into a detailed account of the vexatious delays and official procedures to which he was forced to submit during those twenty-eight long years. It is a matter of satisfaction that Harrison lived to receive the money which he had earned. He had the genius to plan and master intricate mechanical details, but perhaps he lacked in some degree the ability of tongue and pen to bring them home to others. This may be the reason he is so little known, though it was his fortune to contribute so large and essential a part to the perfection of modern navigation. Let us hope this brief mention may serve to recall his memory from oblivion even for a fleeting moment; that we may not have written in vain of that longitude to which his life was given.

THE PLEIADES

Famed in legend; sung by early minstrels of Persia and Hindustan;

"—like a swarm of fire-flies tangled in a silver braid";

yonder distant misty little cloud of Pleiades has always won and held the imagination of men. But it was not only for the inspiration of poets, for quickening fancy into song, that the seven daughters of Atlas were fixed upon the firmament. The problems presented by this group of stars to the unobtrusive scientific investigator are among the most interesting known to astronomy. Their solution is still very incomplete, but what we have already learned may be counted justly among the richest spoils brought back by science from the stored treasure-house of Nature's secrets.

The true student of astronomy is animated by no mere vulgar curiosity to pry into things hidden. If he seeks the concealed springs that move the complex visible mechanism of the heavens, he does so because his imagination is roused by the grandeur of what he sees; and deep down within him stirs the true love of the artist for his art. For it is indeed a fine art, that science of astronomy.

It can have been no mere chance that has massed the Pleiades from among their fellow stars. Men of ordinary eyesight see but a half-dozen distinct objects in the cluster; those of acuter vision can count fourteen; but it is not until we apply the space-penetrating power of the telescope that we realize the extraordinary scale upon which the system of the Pleiades is constructed. With the Paris instrument Wolf in 1876 catalogued 625 stars in the group; and the searching photographic survey of Henry in 1887 revealed no less than 2,326 distinct stars within and near the filmy gauze of nebulous matter always so conspicuous a feature of the Pleiades.

The means at our disposal for the study of stellar distances are but feeble. Only in the case of a very small number of stars have we been able to obtain even so much as an approximate estimate of distance. The most powerful observational machinery, though directed by the tried skill of experience, has not sufficed to sound the profounder depths of space. The Pleiad stars are among those for which no measurement of distance has yet been made, so that we do not know whether they are all equally far away from us. We see them projected on the dark background of the celestial vault; but we cannot tell from actual measurement whether they are all situated near the same point in space. It may be that some are immeasurably closer to us than are the great mass of their companions; possibly we look through the cluster at others far behind it, clinging, as it were, to the very fringe of the visible universe.

Farther on we shall find evidence that something like this really is the case. But under no circumstances is it reasonable to suppose that the whole body of stars can be strung out at all sorts of distances near a straight line pointing in the direction of the visible cluster. Such a distribution may perhaps remain among the possibilities, so long as we cannot measure directly the actual distances of the individual stars. But science never accepts a mere possibility against which we can marshal strong circumstantial evidence. We may conclude on general principles that the gathering of these many objects into a single close assemblage denotes community of origin and interests.

The Pleiades then really belong to one another. What is the nature of their mutual tie? What is their mystery, and can we solve it? The most obvious theory is, of course, suggested by what we know to be true within our own solar system. We owe to Newton the beautiful conception of gravitation, that unique law by means of which astronomers have been enabled to reduce to perfect order the seeming tangle of planetary evolutions. The law really amounts, in effect, to this: All objects suspended within the vacancy of space attract or pull one another. How they can do this without a visible connecting link between them is a mystery which may always remain unsolved. But mystery as it is, we must accept it as an ascertained fact. It is this pull of gravitation that holds together the sun and planets, forcing them all to follow out their due and proper paths, and so to continue throughout an unbroken cycle until the great survivor, Time, shall be no more.

This same gravitational attraction must be at work among the Pleiades. They, too, like ourselves, must have bounds and orbits set and interwoven, revolutions and gyrations far more complex than the solar system knows. The visual discovery of such motion of rotation among the Pleiades may be called one of the pressing problems of astronomy to-day. We feel sure that the time is ripe, and that the discovery is actually being made at the present moment: for a generation of men is not too great a period to call a moment, when we have to deal with cosmic time.

It is indeed the lack of observations extending through sufficient centuries that stays our hand from grasping the coveted result. The Pleiades are so far from us that we cannot be sure of changes among them. Magnitudes are always relative. It matters not how large the actual movements may be; if they are extremely small in comparison with our distance, they must shrink to nothingness in our eyes. Trembling on the verge of invisibility, elusive, they are in that borderland where science as yet but feels her way, though certain that the way is there.

The foundations of exact modern knowledge of the group were laid by Bessel about 1840. With the modesty characteristic of the great, he says quite simply that he has made a number of measures of the Pleiades, thinking that the time may come when astronomers will be able to find some evidence of motion. In this unassuming way he prefaces what is still the classic model of precision and thoroughness in work of this kind. Bessel cleared the ground for a study of inter-stellar motion within the close star-clusters; and it is probable that only by such study may we hope to demonstrate the universality of the law of gravitation in cosmic space.

Bessel's acuteness in forecasting the direction of coming research was amply verified by the work of Elkin in 1885 at Yale College. Provided with a more modern instrument, but similar to Bessel's, Elkin was able to repeat his observations with a slight increase of precision. Motions in the interval of forty-five years, sufficiently great to hint at coming possibilities, were shown conclusively to exist. Six stars at all events have been fairly excluded from the group on account of their peculiar motions shown by Elkin's research. It is possible that they are merely seen in the background through the interstices of the cluster itself, or they may be suspended between us and the Pleiades, in either case having no real connection with the group. Finally, these observations make it reasonably certain that many of the remaining mass of stars really constitute a unit aggregation in space. Astronomers of a coming generation will again repeat the Besselian work. At present we have been able to use his method only for the separation from the true Pleiades of chance stars that happen to lie in the same direction. Let us hope that man shall exist long enough upon this earth to see the clustered stars themselves begin and carry out such gyrations as gravitation imposes.

These will doubtless be of a kind not even suggested by the lesser complexities of our solar system. For the most wonderful thing of all about the Pleiades seems to point to an intricacy of structure whose details may be destined to shake the confidence of the profoundest mathematician. There is an extraordinary nebulous condensation that seems to pervade the entire space occupied by the stellar constituents of the group. The stars are swimming in a veritable sea of luminous cloud. There are filmy tenuous places, and again condensing whirls of material doubtless still in the gaseous or plastic stage. Most noticeable of all are certain almost straight lines of nebula that connect series of stars. In one case, shown upon a photograph made by Henry at Paris, six stars are strung out upon such a hazy line. We might give play to fancy, and see in this the result of some vast eruption of gaseous matter that has already begun to solidify here and there into stellar nuclei. But sound science gives not too great freedom to mere speculative theories. Her duty has been found in quiet research, and her greatest rewards have flowed from imaginative speculation, only when tempered by pure reason.

THE POLE-STAR

One of the most brilliant observations of the last few years is Campbell's recent discovery of the triple character of this star. Centuries and centuries ago, when astronomy, that venerable ancient among the sciences, was but an infant, the pole-star must have been considered the very oldest of observed heavenly bodies. In the beginning it was the only sure guide of the navigator at night, just as to this day it is the foundation-stone for all observational stellar astronomy of precision. There has never been a time in the history of astronomy when the pole-star might not have been called the most frequently measured object in the sky of night. So it is indeed strange that we should find out something altogether new about it after all these ages of study.

But the importance of the discovery rests upon a surer foundation than this. The method by which it has been made is almost a new one in the science. A generation ago, men thought the "perfect science," for so we love to call astronomy, could advance only by increasing a little the exact precision of observation. The citadel of perfect truth might be more closely invested; the forces of science might push forward step by step; the machinery of research might be strengthened, but that a new engine of investigation would be discovered capable of penetrating where no telescope can ever reach, this, indeed, seemed far beyond the liveliest hope of science. Even the discoverer of the spectroscope could never have dreamed of its possibilities, could never have foreseen its successes, its triumphs.

The very name of this instrument suggests mystery to the popular mind. It is set down at once among the things too difficult, too intricate, too abstruse to understand. Yet in its essentials there is nothing about the spectroscope that cannot be made clear in a few words. Even the modern "undulatory theory" of light itself is terrible only in the length of its name. Anyone who has seen the waves of ocean roll, roll, and ever again roll in upon the shore, can form a very good notion of how light moves. 'Tis just such a series of rolling waves; started perhaps from some brilliant constellation far out upon the confining bounds of the visible universe, or perhaps coming from a humble light upon the student's table; yet it is never anything but a succession of rolling waves. Only, unlike the waves of the sea, light waves are all excessively small. We should call one whose length was a twenty-thousandth of an inch a big one!

Now the human eye possesses the property of receiving and understanding these little waves. The process is an unconscious one. Let but a set of these tiny waves roll up, as it were, out of the vast ocean of space and impinge upon the eye, and all the phenomena of light and color become what we call "visible." We see the light.

And how does all this find an application in astronomy? Not to enter too much into technical details, we may say that the spectroscope is an instrument which enables us to measure the length of these light waves, though their length is so exceedingly small. The day has indeed gone by when that which poets love to call the Book of Nature was printed in type that could be read by the eye unaided. Telescope, microscope, and spectroscope are essential now to him who would penetrate any of Nature's secrets. But measurements with a telescope, like eye observations, are limited strictly to determining the directions in which we see the heavenly bodies. Ever since the beginning of things, when old Hipparchus and Ulugh Beg made the first rude but successful attempts to catalogue the stars, the eye and telescope have been able to measure only such directions. We aim the telescope at a star, and record the direction in which it was pointed. Distances in astronomy can never be measured directly. All that we know of them has been obtained by calculations based upon the

Newtonian law of gravitation and observations of directions.

Now the spectroscope seems to offer a sort of exception to this rule. Suppose we can measure the wave-lengths of the light sent us from a star. Suppose again that the star is itself moving swiftly toward us through space, while continually setting in motion the waves of light that are ultimately to reach the waiting astronomer. Evidently the light waves will be crowded together somewhat on account of the star's motion. More waves per second will reach us than would be received from a star at rest. It is as though the light waves were compressed or shortened a little. And if the star is leaving us, instead of coming nearer, opposite effects will occur. We have then but to compare spectroscopically starlight with some artificial source of light in the observatory in order to find out whether the star is approaching us or receding from us. And by a simple process of calculation this stellar motion can be obtained in miles per second. Thus we can now actually measure directly, in a certain sense, linear speed in stellar space, though we are still without the means of getting directly at stellar distances.

But the most wonderful thing of all about these spectroscopic measures is the fact that it makes no difference whatever how far away is the star under observation. What we learn through the spectroscope comes from a study of the waves themselves, and it is of no consequence how far they have travelled, or how long they have been a-coming. For it must not be supposed that these waves consume no time in passing from a distant star to our own solar system. It is true that they move exceeding fast; certainly 180,000 miles per second may be called rapid motion. But if this cosmic velocity of light is tremendous, so also are cosmic distances correspondingly vast. Light needs to move quickly coming from a star, for even at the rate of motion we have mentioned it requires many years to reach us from some of the more distant constellations. It has been well said that an observer on some far-away star, if endowed with the power to see at any distance, however great, might at this moment be looking on the Crusaders proceeding from Europe against the Saracen at Jerusalem. For it is quite possible that not until now has the light which would make the earth visible had time to reach him. Yet distant as such an observer might be, light from the star on which he stood could be measured in the spectroscope, and would infallibly tell us whether the earth and star are approaching in space or gradually drawing farther asunder.

The pole-star is not one of the more distant stellar systems. We do not know how far it is from us very exactly, but certainly not less than forty or fifty years are necessary for its light to reach us. The star might have gone out of existence twenty years ago, and we not yet know of it, for we would still be receiving the light which began its long journey to us about 1850 or 1860. But no matter what may be its distance, Campbell found by careful observations, made in the latter part of 1896, that the pole-star was then approaching the earth at the rate of about twelve miles per second. So far there was nothing especially remarkable. But in August and September of the present year twenty-six careful determinations were made, and these showed that now the rate of approach varied between about five and nine miles per second. More astonishing still, there was a uniform period in the changes of velocity. In about four days the rate of motion changed from about five to nine miles and back again. And this variation kept on with great regularity. Every successive period of four days saw a complete cycle of velocity change forward and back between the same limits. There can be but one reasonable explanation. This star must be a double, or "binary" star. The two components, under the influence of powerful mutual gravitational attraction, must be revolving in a mighty orbit. Yet this vast orbit, as a whole, with the two great stars in it, must be approaching our part of the universe all the time. For the spectroscope shows the velocity of approach to increase and diminish, indeed, but it is always present. Here, then, is this great stellar system, having a four-day revolution of its own, and yet swinging rapidly through space in our direction. Nor is this all. One of the component stars must be nearly or quite dark; else its presence would infallibly be detected by our instruments.

And now we come to the most astonishing thing of all. How comes it that the average rate of approach of the "four-day system," as a whole, changed between 1896 and 1899? In 1896 only this velocity of the whole system was determined, the four-day period remaining undiscovered until the more numerous observations of 1899. But even without considering the four-day period, the changing velocity of the entire system offers one of those problems that exact science can treat only by the help of the imagination. There must be some other great centre of attraction, some cosmic giant, holding the visible double pole-star under its control. Thus, that which we see, and call the pole-star, is in reality threading its path about the third and greatest member of the system, itself situated in space, we know not where.

Spiral Nebula in Constellation Leo.
Photographed by Keeler, February 24, 1900.
Exposure, three hours, fifty minutes.

NEBULÆ

Scattered about here and there among the stars are certain patches of faint luminosity called by astronomers Nebulæ. These "little clouds" of filmy light are among the most fascinating of all the kaleidoscopic phenomena of the heavens; for it needs but a glance at one of them to give the impression that here before us is the stuff of which worlds are made. All our knowledge of Nature leads us to expect in her finished work the result of a series of gradual processes of development. Highly organized phenomena such as those existing in our solar system did not spring into perfection in an instant. Influential forces, easy to imagine, but difficult to define, must have directed the slow, sure transformation of elemental matter into sun and planets, things and men. Therefore a study of those forces and of their probable action upon nebular material has always exerted a strong attraction upon the acutest thinkers among men of exact science.

Our knowledge of the nebulæ is of two kinds—that which has been ascertained from observation as to their appearance, size, distribution, and distance; and that which is based upon hypotheses and theoretical reasoning about the condensation of stellar systems out of nebular masses. It so happens that our observational material has received a very important addition quite recently through the application of photography to the delineation of nebulæ, and this we shall describe farther on.

Two nebulæ only are visible to the unaided eye. The brighter of these is in the constellation Andromeda; it is of oval or elliptical shape, and has a distinct central condensation or nucleus. Upon a photograph by Roberts it appears to have several concentric rings surrounding the nebula proper, and gives the general impression of a flat round disk foreshortened into an oval shape on account of the observer's position not being square to the surface of the disk. Very recent photographs of this nebula, made with the three-foot reflecting telescope of the Lick Observatory, bring out the fact that it is really spiral in form, and that the outlying nebulous rings are only parts of the spires in a great cosmic whorl.

Nebula in Andromeda.
Lower object in the photograph is a Comet.
Photographed by Barnard, November 21, 1892.

This Andromeda nebula is the one in which the temporary star of 1885 appeared. It blazed up quite suddenly near the apparent centre of the nebula, and continued in view for six months, fading finally beyond the reach of our most powerful telescopes. There can be little doubt that the star was actually in the nebula, and not merely seen through it, though in reality situated in the extreme outlying part of space at a distance immeasurably greater than that separating us from the nebula itself. Such an accidental superposition of nebula and star might even be due to sudden incandescence of a new star between us and the nebula. In such a case we should see the star projected upon the surface of the nebula, so that the superposition would be identical with that actually observed. Therefore, while it is, indeed, possible that the star may have been either far behind the nebula or in front of it, we must accept as more probable the supposition that there was a real connection between the two. In that case there is little doubt that we have actually observed one of those cataclysms that mark successive steps of cosmic evolution. We have no thoroughly satisfactory theory to account for such an explosive catastrophe within the body of the nebula itself.

The other naked-eye nebula is in the constellation Orion. In the telescope it is a more striking object, perhaps, than the Andromeda nebula; for it has no well-defined geometrical form, but consists of an immense odd-shaped mass of light enclosing and surrounding a number of stars. It is unquestionably of a very complicated structure, and is, therefore, less easily studied and explained than the nebulæ of simpler form. There is no doubt that the Orion nebula is composed of luminous gas, and is not merely a cluster of small stars too numerous and too near together to be separated from each other, even in our most powerful telescopes. It was, indeed, supposed, until about forty years ago, that all the nebulæ are simply irresolvable star-clusters; but we now have indisputable evidence, derived from the spectroscope, that many nebulæ are composed of true gases, similar to those with which we experiment in chemical laboratories. This spectroscopic proof of the gaseous character of nebulæ is one of the most important discoveries contributed by that instrument to our small stock of facts concerning the structure of the sidereal universe.

Coming now to the smaller nebulæ, we find a great diversity of form and appearance. Some are ring-shaped, perhaps having a less brilliant nebulosity within the ring. Many show a central condensation of disk-like appearance (planetary nebulæ), or have simply a star at the centre (nebulous stars). Altogether about ten thousand such objects have been catalogued by successive generations of astronomers since the invention of the telescope, and most of these have been reported as oval in form. Now we have already referred to the important addition to our knowledge of the nebulæ obtained by recent photographic observations; and this addition consists in the discovery that most of these oval nebulæ are in reality spirals. Indeed, it appears that the spiral type is the normal type, and that nebulæ of irregular or other forms are exceptions to the general rule. Even the great Andromeda nebula, as we have seen, is now recognized as a spiral.

The instrument with which its convolute structure was discovered is a three-foot reflecting telescope, made by Common of England, and now mounted at the Lick Observatory, in California. The late Professor Keeler devoted much of his time to photographing nebulæ during the last year or two. He was able to establish the important fact just mentioned, that most nebulæ formerly thought to be mere ovals, turn out to be spiral when brought under the more searching scrutiny of the photographic plate applied at the focus of a telescope of great size, and with an exposure to the feeble nebular light extending through three or four consecutive hours.

Many of the spirals have more than a single volute. It is as though one were to attach a number of very flexible rods to an axle, like spokes of a wheel without a rim and then revolve the axle rapidly. The flexible rods would bend under the rapid rotation, and form a series of spiral curves not unlike many of these nebulæ. Indeed, it is impossible to escape the conviction that these great celestial whorls are whirling around an axis. And it is most important in the study of the growth of worlds, to recognize that the type specimen is a revolving spiral. Therefore, the rotating flattened globe of incandescent matter postulated by Laplace's nebular hypothesis would make of our solar system an exceptional world, and not a type of stellar evolution in general.

Keeler's photographs have taught us one thing more. Scarcely is there a single one of his negatives that does not show nebulæ previously uncatalogued. It is estimated that if this process of photography could be extended so as to cover the entire sky, the whole number of nebulæ would add up to the stupendous total of 120,000; and of these the great majority would be spiral.

When we approach the question of the distribution of nebulæ in different parts of the sky, as shown by their catalogued positions, we are met by a curious fact. It appears that the region in the neighborhood of the Milky Way is especially poor in nebulæ, whereas these objects seem to cluster in much larger numbers about those points in the sky that are farthest from the Milky Way. But we know that the Milky Way is richer in stars than any other part of the sky, since it is, in fact, made up of stellar bodies clustered so closely that it is wellnigh impossible to see between them in the denser portions. Now, it cannot be the result of chance that the stars should tend to congregate in the Milky Way, while the nebulæ tend to seek a position as far from it as possible. Whatever may be the cause, we must conclude that the sidereal system, as we see it, is in general constructed upon a single plan, and does not consist of a series of universes scattered at random throughout space. If we are to suppose that nebulæ turn into stars as a result of condensation or any other change, then it is not astonishing to find a minimum of nebulæ where there is a maximum of stars, since the nebulæ will have been consumed, as it were, in the formation of the stars.

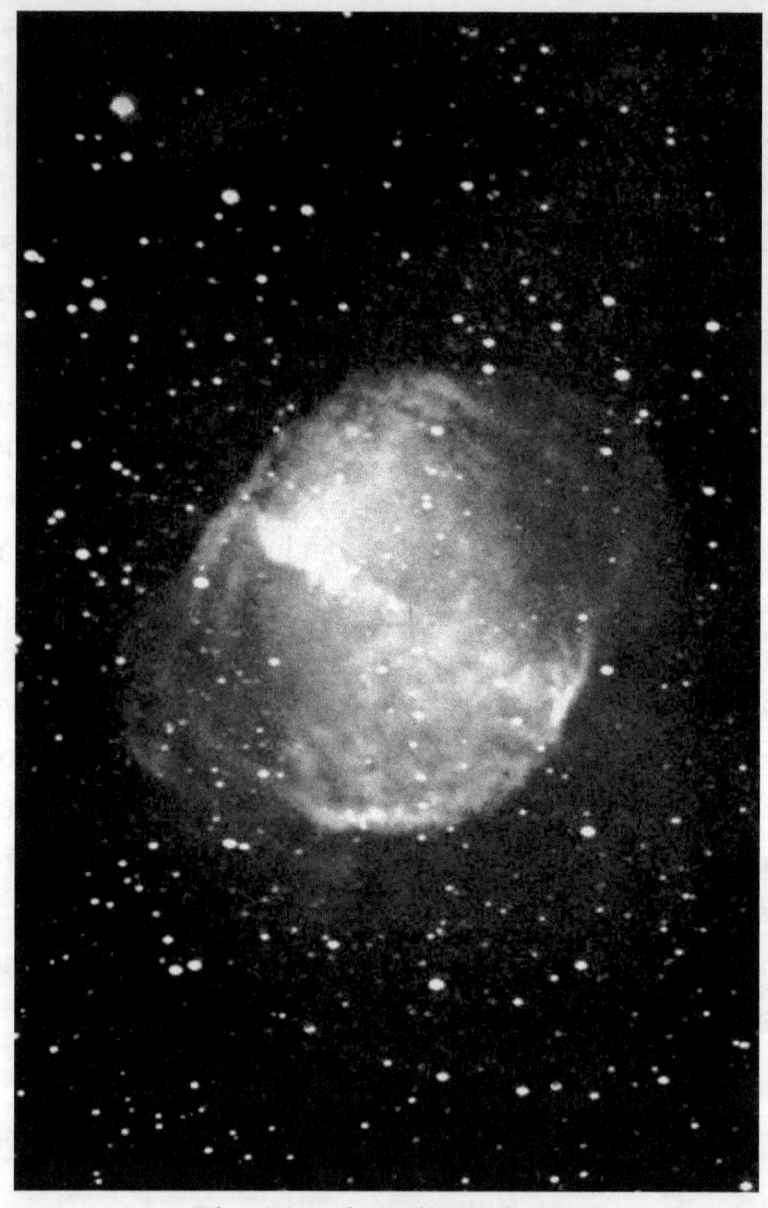

The "Dumb-Bell" Nebula.
Photographed by Keeler, July 31, 1899.
Exposure, three hours.

It is never advisable to push philosophical speculation very far when supported by too slender a basis of fact. But if we are to regard the visible universe as made up on the whole of a single system of bodies, we may well ask one or two questions to be answered by speculative theory. We have said the stars are not uniformly distributed in space. Their concentration in the Milky Way, forming a narrow band dividing the sky into two very nearly equal parts, must be due to their being actually massed in a thin disk or ring of space within which our solar system is also situated. This thin disk projected upon the sky would then appear as the narrow star-band of the Milky Way. Now, suppose this disk has an axis perpendicular to itself, and let us imagine a rotation of the whole sidereal system about that axis. Then the fact that the visible nebulæ are congregated far from the Milky Way means that they are actually near the imaginary axis.

Possibly the diminished velocity of motion near the axis may have something to do with the presence of the nebulæ there. Possibly the nebulæ themselves have axes perpendicular to the plane of the Milky Way. If so, we should see the spiral nebulæ near the Milky Way edgewise, and those far from it without foreshortening. Thus, the paucity of nebulæ near the Milky Way may be due in part to the increased difficulty of seeing them when looked at edgewise. Indeed, there is no limit to the possibilities of hypothetical reasoning about the nebular structure of our universe; unfortunately, the whole question must be placed for the present among those intensely interesting cosmic problems awaiting elucidation, let us hope, in this new century.

TEMPORARY STARS

Nothing can be more erroneous than to suppose that the stellar multitude has continued unchanged throughout all generations of men. "Eternal fires" poets have called the stars; yet they burn like any little conflagration on the earth; now flashing with energy, brilliant, incandescent, and again sinking into the dull glow of smouldering half-burned ashes. It is even probable that space contains many darkened orbs, stars that may have risen in constellations to adorn the skies of prehistoric time—now cold, unseen, unknown. So far from dealing with an unvarying universe, it is safe to say that sidereal astronomy can advance only by the discovery of change. Observational science watches with untiring industry, and night hides few celestial events from the ardent scrutiny of astronomers. Old theories are tested and newer ones often perfected by the detection of some slight and previously unsuspected alteration upon the face of the sky. The interpretation of such changes is the most difficult task of science; it has taxed the acutest intellects among men throughout all time.

If, then, changes can be seen among the stars, what are we to think of the most important change of all, the blazing into life of a new stellar system? Fifteen times since men began to write their records of the skies has the birth of a star been seen. Surely we may use this term when we speak of the sudden appearance of a brilliant luminary where nothing visible existed before. But we shall see further on that scientific considerations make it highly probable that the phenomenon in question does not really involve the creation of new matter. It is old material becoming suddenly luminous for some hidden reason. In fact, whenever a new object of great brilliancy has been discovered, it has been found to lose its light again quite soon, ending either in total extinction or at least in comparative darkness. It is for this reason that the name "temporary star" has been applied to cases of this kind.

The first authenticated instance dates from the year 134 B.C., when a new star appeared in the constellation Scorpio. It was this star that led Hipparchus to construct his stellar catalogue, the first ever made. It occurred to him, of course, that there could be but one way to make sure in the future that any given object discovered in the sky was new; it was necessary to make a complete list of everything visible in his day. Later astronomers need then only compare Hipparchus's catalogue with the heavens from time to time in order to find out whether anything unknown had appeared. This work of

Hipparchus became the foundation of sidereal study, and led to most important discoveries of various kinds.

But no records remain concerning his new star except the bare fact of its appearance in Scorpio. Hipparchus's published works are all lost. We do not even know the exact place of his birth, and as for those two dates of entry and exit that history attaches to great names—we have them not. Yet he was easily the first astronomer of antiquity, one of the first of all time; and we know of him only from the writings of Ptolemy, who lived three hundred years after him.

More than five centuries elapsed before another temporary star was entered in the records of astronomy. This happened in the year 389 A.D., when a star appeared in Aquila; and of this one also we know nothing further. But about twelve centuries later, in November, 1572, a new and brilliant object was found in the constellation Cassiopeia. It is known as Tycho's star, since it was the means of winning for astronomy a man who will always take high rank in her annals, Tycho Brahe, of Denmark. When he first saw this star, it was already very bright, equalling even Venus at her best; and he continued a careful series of observations for sixteen months, when it faded finally from his view. The position of the new star was measured with reference to other stars in the constellation Cassiopeia, and the results of Tycho's observations were finally published by him in the year 1573. It appears that much urging on the part of friends was necessary to induce him to consent to this publication, not because of a modest reluctance to rush into print, but for the reason that he considered it undignified for a nobleman of Denmark to be the author of a book!

An important question in cosmic astronomy is opened by Tycho's star. Did it really disappear from the heavens when he saw it no more, or had its lustre simply been reduced below the visual power of the unaided eye? Unfortunately, Tycho's observations of the star's position in the constellation were necessarily crude. He possessed no instruments of precision such as we now have at our disposal, and so his work gives us only a rather rough approximation of the true place of the star. A small circle might be imagined on the sky of a size comparable with the possible errors of Tycho's observations. We could then say with certainty that his star must have been situated somewhere within that little circle, but it is impossible to know exactly where.

It happens that our modern telescopes reveal the existence of several faint stars within the space covered by such a circle. Any one of these would have been too small for Tycho to see, and, therefore, any one of them may be his once brilliant luminary reduced to a state of permanent or temporary semi-darkness. These considerations are, indeed, of great importance in explaining the phenomena of temporary stars. If Tycho had been able to leave us a more exact determination of his star's place in the sky, and even if our most powerful instruments could not show anything in that place to-day, we might nevertheless theorize on the supposition that the object still exists, but has reached a condition almost entirely dark.

Indeed, the latest theory classes temporary stars among those known as variable. For many stars are known to undergo quite decided changes in brilliancy; possibly inconstancy of light is the rule rather than the exception. But while such changes, when they exist, are too small to be perceptible in most cases, there is certainly a large number of observable variables, subject to easily measurable alterations of light. Astronomers prefer to see in the phenomena of temporary stars simple cases of variation in which the increase of light is sudden, and followed by a gradual diminution. Possibly there is then a long period of comparative or even complete darkness, to be followed as before by a sudden blazing up and extinction. No temporary star, however, has been observed to reappear in the same celestial place where once had glowed its sudden outburst. But cases are not wanting where incandescence has been both preceded and followed by a continued existence, visible though not brilliant.

For such cases as these it is necessary to come down to modern records. We cannot be sure that some faint star has been temporarily brilliant, unless we actually see the conflagration itself, or are able to make the identity of the object's precise location in the sky before and after the event perfectly certain by the aid of modern instruments of precision. But no one has ever seen the smouldering fires break out. Temporary stars have always been first noticed only after having been active for hours if not for days. So we must perforce fall back on instrumental identification by determinations of the star's exact position upon the celestial vault.

Some time between May 10th and 12th in the year 1866 the ninth star in the list of known "temporaries" appeared. It possessed very great light-giving power, being surpassed in brilliancy by only about a score of stars in all the heavens. It retained a maximum luminosity only three or four days, and in less than two months had diminished to a point somewhere between the ninth and tenth "magnitudes." In other words, from a conspicuous star, visible to the naked eye, it had passed beyond the power of anything less than a good telescope. Fortunately, we had excellent star-catalogues before 1866. These were at once searched, and it was possible to settle quite definitely that a star of about the ninth or tenth magnitude had really existed before 1866 at precisely the same point occupied by the new one. Needless to say, observations were made of the new star itself, and afterward compared with later observations of the faint one that still occupies its place. These render quite certain the identity of the temporary bright star with the faint ones that preceded and followed it.

Such results, on the one hand, offer an excellent vindication of the painstaking labor expended on the construction of star-catalogues, and, on the other, serve to elucidate the mystery of temporary stars. Nothing can be more plausible than to explain by analogy those cases in which no previous or subsequent existence has been observed. It is merely necessary to suppose that, instead of varying from the ninth or tenth magnitude, other temporary objects have begun and ended with the twentieth; for the twentieth magnitude would be beyond the power of our best instruments.

Nor is the star of 1866 an isolated instance. Ten years later, in 1876, a temporary star blazed up to about the second magnitude, and returned to invisibility, so far as the naked eye is concerned, within a month, having retained its greatest brilliancy only one or two days. This star is still visible as a tiny point of light, estimated to be of the fifteenth magnitude. Whether it existed prior to its sudden outburst can never be known, because we do not possess catalogues including the generality of stars as faint as this one must have been. But at all events, the continued existence of the object helps to place the temporary stars in the class of variables.

The next star, already mentioned under "nebula," was first seen in 1885. It was in one respect the most remarkable of all, for it appeared almost in the centre of the great nebula in the constellation Andromeda. It was never very bright, reaching only the sixth magnitude or thereabouts, was observed during a period of only six months, and at the end of that time had faded beyond the reach of our most powerful glasses. It is a most impressive fact that this event occurred within the nebula. Whatever may be the nature of the explosive catastrophe to which the temporary stars owe their origin, we can now say with certainty that not even those vast elemental luminous clouds men call nebulæ are free from danger.

The last outburst on our records was first noticed February 22, 1901. The star appeared in the constellation Perseus, and soon reached the first magnitude, surpassing almost every other star in the sky. It has been especially remarkable in that it has become surrounded by a nebulous mass in which are several bright condensations or nuclei; and these seem to be in very rapid motion. The star is still under observation (January, 1902).

GALILEO

Among the figures that stand out sharply upon the dim background of old-time science, there is none that excites a keener interest than Galileo. Most people know him only as a distinguished man of learning; one who carried on a vigorous controversy with the Church on matters scientific. It requires some little study, some careful reading between the lines of astronomical history, to gain acquaintance with the man himself. He had a brilliant, incisive wit; was a genuine humorist; knew well and loved the amusing side of things; and could not often forego a sarcastic pleasantry, or deny himself the pleasure of argument. Yet it is more than doubtful if he ever intended impertinence, or gave willingly any cause of quarrel to the Church.

His acute understanding must have seen that there exists no real conflict between science and religion; for time, in passing, has made common knowledge of this truth, as it has of many things once hidden. When we consider events that occurred three centuries ago, it is easy to replace excited argument with cool judgment; to remember that those were days of violence and cruelty; that public ignorance was of a density difficult to imagine to-day; and that it was universally considered the duty of the Church to assume an authoritative attitude upon many questions with which she is not now required to concern herself in the least. Charlatans, unbalanced theorists, purveyors of scientific marvels, were all liable to be passed upon definitely by the Church, not in a spirit of impertinent interference, but simply as part of her regular duties.

If the Church's judgment in such matters was sometimes erroneous; if her interference now and again was cruel, the cause must be sought in the manners and customs of the time, when persecution rioted in company with ignorance, and violence was the law. Perhaps even to-day it would not be amiss to have a modern scientific board pass authoritatively upon novel discoveries and inventions, so as to protect the public against impostors as the Church tried to do of old.

Galileo was born at Pisa in 1564, and his long life lasted until 1642, the very year of Newton's birth. His most important scientific discoveries may be summed up in a few words; he was the first to use a telescope for examining the heavenly bodies; he discovered mountains on the moon; the satellites of Jupiter; the peculiar appearance of Saturn which Huygens afterward explained as a ring surrounding the ball of the planet; and, finally, he found black spots on the sun's disk. These discoveries, together with his remarkable researches in mechanical science, constitute Galileo's claim to immortality as an investigator. But, as we have said, it is not our intention to consider his work as a series of scientific discoveries. We shall take a more interesting point of view, and deal with him rather as a human being who had contracted the habit of making scientific researches.

What must have been his feelings when he first found with his "new" telescope the satellites of Jupiter? They were seen on the night of January 7, 1610. He had already viewed the planet through his earlier and less powerful glass, and was aware that it possessed a round disk like the moon, only smaller. Now he saw also three objects that he took to be little stars near the planet. But on the following night, as he says, "drawn by what fate I know not," the tube was again turned upon the planet. The three small stars had changed their positions, and were now all situated to the west of Jupiter, whereas on the previous night two had been on the eastern side. He could not explain this phenomenon, but he recognized that there was something peculiar at work. Long afterward, in one of his later works, translated into quaint old English by Salusbury, he declared that "one sole experiment sufficeth to batter to the ground a thousand probable

Arguments." This was already the guiding principle of his scientific activity, a principle of incomparable importance, and generally credited to Bacon. Needless to say, Jupiter was now examined every night.

The 9th was cloudy, but on the 10th he again saw his little stars, their number now reduced to two. He guessed that the third was behind the planet's disk. The position of the two visible ones was altogether different from either of the previous observations. On the 11th he became sure that what he saw was really a series of satellites accompanying Jupiter on his journey through space, and at the same time revolving around him. On the 12th, at 3 A.M., he actually saw one of the small objects emerge from behind the planet; and on the 13th he finally saw four satellites. Two hundred and eighty-two years were destined to pass away before any human eye should see a fifth. It was Barnard in 1892 who followed Galileo.

To understand the effect of this discovery upon Galileo requires a person who has himself watched the stars, not, as a dilettante, seeking recreation or amusement, but with that deep reverence that comes only to him who feels—nay, knows—that in the moment of observation just passed he too has added his mite to the great fund of human knowledge. Galileo's mummied forefinger still points toward the stars from its little pedestal of wood in the *Museo* at

Florence, a sign to all men that he is unforgotten. But Galileo knew on that 11th of January, 1610, that the memory of him would never fade; that the very music of the spheres would thenceforward be attuned to a truer note, if any would but hearken to the Jovian harmony. For he recognized at once that the visible revolution of these moons around Jupiter, while that planet was himself visibly travelling through space, must deal its death-blow to the old Ptolemaic system of the universe. Here was a great planet, the centre of a system of satellites, and yet not the centre of the universe. Surely, then, the earth, too, might be a mere planet like Jupiter, and not the supposed motionless centre of all things.

The satellite discovery was published in 1610 in a little book called "Sidereus Nuncius," usually translated "The Sidereal Messenger." It seems to us, however, that the word "messenger" is not strong enough; surely in Papal Italy a *nuncius* was more than a mere messenger. He was clothed with the very highest authority, and we think it probable that Galileo's choice of this word in the title of his book means that he claimed for himself similar authority in science. At all events, the book made him at once a great reputation and numerous enemies.

But it was not until 1616 that the Holy Office (Inquisition) issued an edict ordering Galileo to abandon his opinion that the earth moved, and at the same time placed Copernicus's *De Revolutionibus* and two other books advocating that doctrine on the "Index Librorum Prohibitorum," or list of books forbidden by the Church. These volumes remained in subsequent editions of the "Index" down to 1821, but they no longer appear in the edition in force to-day.

Galileo's most characteristic work is entitled the "Dialogue on the Two Chief Systems of the World." It was not published until 1632, although the idea of the book was conceived many years earlier. In it he gave full play to his extraordinary powers as a true humorist, a *fine lame* among controversialists, and a genuine man of science, valuing naked truth above all other things. As may be imagined, it was no small matter to obtain the authorities' consent to this publication. Galileo was already known to hold heretical opinions, and it was suspected that he had not laid them aside when commanded to do so by the edict of 1616. But perhaps Galileo's introduction to the "Dialogue" secured the censor's *imprimatur*; it is even suspected that the Roman authorities helped in the preparation of this introduction. Fortunately, we have a delightful contemporary translation into English, by Thomas Salusbury, printed at London by Leybourne in 1661. We have already quoted from this translation, and now add from the same work part of Galileo's masterly preface to the "Dialogue":

"Judicious reader, there was published some years since in *Rome* a salutiferous Edict, that, for the obviating of the dangerous Scandals of the Present Age, imposed a reasonable Silence upon the Pythagorean (Copernican) opinion of the Mobility of the Earth. There want not such as unadvisedly affirm, that the Decree was not the production of a sober Scrutiny, but of an ill-formed passion; and one may hear some mutter that Consultors altogether ignorant of Astronomical observations ought not to clipp the wings of speculative wits with rash prohibitions."

Galileo first states his own views, and then pretends that he will oppose them. He goes on to say that he believes in the earth's immobility, and takes "the contrary only for a mathematical *Capriccio*," as he calls it; something to be considered, because possessing an academical interest, but on no account having a real existence. Of course any one (even a censor) ought to be able to see that it is the Capriccio, and not its opposite, that Galileo really advocates. Three persons appear in the "Dialogue": Salviati, who believes in the Copernican system; Simplicio, of suggestive name, who thinks the earth cannot move; and, finally, Sagredus, a neutral gentleman of humorous propensities, who usually begins by opposing Salviati, but ends by being convinced. He then helps to punish poor Simplicio, who is one of those persons apparently incapable of comprehending a reasonable argument. Here is an interesting specimen of the "Dialogue" taken from Salusbury's translation: Salviati refers to the argument, then well known, that the earth cannot rotate on its axis, "because of the impossibility of its moving long without wearinesse." Sagredus replies: "There are some kinds of animals which refresh themselves after wearinesse by rowling on the earth; and that therefore there is no need to fear that the Terrestrial Globe should tire, nay, it may be reasonably affirmed that it enjoyeth a perpetual and most tranquil repose, keeping itself in an eternal rowling." Salviati's comment on this sally is, "You are too tart and satyrical, Sagredus."

There is no doubt that the "Dialogue" finished the Ptolemaic theory, and made that of Copernicus the only possible one. At all events, it brought about the well-known attack upon Galileo from the authorities of the Holy Office. We shall not recount the often-told tale of his recantation. He was convicted (very rightly) of being a Copernican, and was forced to abjure that doctrine. Galileo's life may be summed up as one of those through which the world has been made richer. A clean-cutting analytic wit, never becoming dull:

heated again and again in the fierce blaze of controversy, it was allowed to cool only that it might acquire a finer temper, to pierce with fatal certainty the smallest imperfections in the armor of his adversaries.

THE PLANET OF 1898

The discovery of a new and important planet usually receives more immediate popular attention and applause than any other astronomical event. Philosophers are fond of referring to our solar system as a mere atom among the countless universes that seem to be suspended within the profound depths of space. They are wont to point out that this solar system, small and insignificant as a whole in comparison with many of the stellar worlds, is, nevertheless, made up of a large number of constituent planets; and these in turn are often accompanied with still smaller satellites, or moons. Thus does Nature provide worlds within worlds, and it is not surprising that public attention should be at once attracted by any new member of our sun's own special family of planets. The ancients were acquainted with only five of the bodies now counted as planets, viz.: Mercury, Venus, Mars, Jupiter, and Saturn. The dates of their discovery are lost in antiquity. To these Uranus was added in 1781 by a brilliant effort of the elder Herschel. We are told that intense popular excitement followed the announcement of Herschel's first observation: he was knighted and otherwise honored by the English King, and was enabled to lay a secure foundation for the future distinguished astronomical reputation of his family.

Herschel's discovery quickened the restless activity of astronomers. Persistent efforts were made to sift the heavens more and more closely, with the strengthened hope of adding still further to our planetary knowledge. An association of twenty-four enthusiastic German astronomers was formed for the express purpose of hunting planets. But it fell to the lot of an Italian, Piazzi, of Palermo, to find the first of that series of small bodies now known as the asteroids or minor planets. He made the discovery at the very beginning of our century, January 1, 1801.

But news travelled slowly in those days, and it was not until nearly April that the German observers heard from Piazzi. In the meantime, he had himself been prevented by illness from continuing his observations. Unfortunately, the planet had by this time moved so near the sun, on account of its own motions and those of the earth, that it could no longer be observed. The bright light of the sun made observations of the new body impossible; and it was feared that, owing to lack of knowledge of the planet's orbit, astronomers would be unable to trace it. So there seemed, indeed, to be danger of an almost irreparable loss to science. But in scientific, as in other human emergencies, someone always appears at the proper moment. A very young mathematician at Göttingen, named Gauss, attacked the problem, and was able to devise a method of predicting the future course of the planet on the sky, using only the few observations made by Piazzi himself. Up to that time no one had attempted to compute a planetary orbit, unless he had at his disposal a series of observations extending throughout the whole period of the planet's revolution around the sun. But the Piazzi planet offered a new problem in astronomy. It had become imperatively necessary to obtain an orbit from a few observations made at nearly the same date. Gauss's work was signally triumphant, for the planet was actually found in the position predicted by him, as soon as a change in the relative places of the planet and earth permitted suitable observations to be made.

But after all, Piazzi's planet belongs to a class of quite small bodies, and is by no means as interesting as Herschel's discovery, Uranus. Yet even this must be relegated to second rank among planetary discoveries. On September 23, 1846, the telescope of the Berlin Observatory was directed to a certain point on the sky for a very special reason. Galle, the astronomer of Berlin, had received a letter from Leverrier, of Paris, telling him that if he would look in a certain direction he would detect a new and large planet.

Leverrier's information was based upon a mathematical calculation. Seated in his study, with no instruments but pen and paper, he had slowly figured out the history of a world as yet unseen. Tiny discrepancies existed in the observed motions of Herschel's planet Uranus. No man had explained their cause. To Leverrier's acute understanding they slowly shaped themselves into the possible effects of attraction emanating from some unknown planet exterior to Uranus. Was it conceivable that these slight tremulous imperfections in the motion of a planet could be explained in this way? Leverrier was able to say confidently, "Yes." But we may rest assured that Galle had but small hopes that upon his eye first, of all the myriad eyes of men, would fall a ray of the new planet's light. Careful and methodical, he would neglect no chance of advancing his beloved science. He would look.

Only one who has himself often seen the morning's sunrise put an end to a night's observation of the stars can hope to appreciate what Galle's feelings must have been when he saw the planet. To his trained eye it was certainly recognizable at once. And then the good news was sent on to Paris. We can imagine Leverrier, the cool calculator, saying to himself: "Of course he found it. It was a mathematical certainty." Nevertheless, his satisfaction must have been of the keenest. No triumphs give a pleasure higher than those of the intellect. Let no one imagine that men who make researches in the domain of pure science are under-paid. They find their reward in pleasure that is beyond any price.

The Leverrier planet was found to be the last of the so-called major planets, so far as we can say in the present state of science. It received the name Neptune. Observers have found no other member of the solar system comparable in size with such bodies as Uranus and Neptune. More than one eager mathematician has tried to repeat Leverrier's achievement, but the supposed planet was not found. It has been said that figures never lie; yet such is the case only when the computations are correctly made. People are prone to give to the work of careless or incompetent mathematicians the same degree of credence that is really due only to masters of the craft. It requires the test of time to affix to any man's work the stamp of true genius.

While, then, we have found no more large planets, quite a group of companions to Piazzi's little one have been discovered. They are all small, probably never exceeding about 400 miles in diameter. All travel around the sun in orbits that lie wholly within that of Jupiter and are exterior to that of Mars. The introduction of astronomical photography has given a tremendous impetus to the discovery of these minor planets, as they are called. It is quite interesting to examine the photographic process by which such discoveries are made possible and even easy. The matter will not be difficult to understand if we remember that all the planets are continually changing their places among the other stars. For the planets travel around the sun at a comparatively small distance. The great majority of the stars, on the contrary, are separated from the sun by an almost immeasurable space. As a result, they do not seem to move at all among themselves, and so we call them fixed stars: they may, indeed, be in motion, but their great distance prevents our detecting it in a short period of time.

Now, stellar photographs are made in much the same way as ordinary portraits. Only, instead of using a simple camera, the astronomer exposes his photographic plate at the eye-end of a telescope. The sensitive surface of the plate is substituted for the human eye. We then find on the picture a little dot corresponding to every star within the photographed region of the sky. But, as everyone knows, the turning of the earth on its axis makes the whole heavens, including the sun, moon, and stars, rise and set every day. So the stars, when we photograph them, are sure to be either climbing up in the eastern sky or else slowly creeping down in the western. And that makes astronomical photography very different from ordinary portrait work.

The stars correspond to the sitter, but they don't sit still. For this reason it is necessary to connect the telescope with a mechanical contrivance which makes it turn round like the hour-hand of an ordinary clock. The arrangement is so adjusted that the telescope, once aimed at the proper object in the sky, will move so as to remain pointed exactly the same during the whole time of the photographic exposure. Thus, while the light of any star is acting on the plate, such action will be continuous at a single point. Consequently, the finished picture will show the star as a little dot; while without this arrangement, the star would trail out into a line instead of a dot. Now we have seen that the planets are all moving slowly among the fixed stars. So if we make a star photograph in a part of the sky where a planet happens to be, the planet will make a short line on the plate; whereas, if the planet remained quite unmoved relatively to the stars it would give a dot like the star dots. The presence of a line, therefore, at once indicates a planet.

This method of planet-hunting has proved most useful. More than 400 small planets similar to Piazzi's have been found, though never another one like Uranus and Neptune. As we have said, all these little bodies lie between Mars and Jupiter. They evidently belong to a group or family, and many astronomers have been led to believe that they are but fragments of a former large planet.

In August, 1898, however, one was found by Witt, of Berlin, which will probably occupy a very prominent place in the annals of astronomy. For this planet goes well within the orbit of Mars, and this will bring it at times very close to the earth. In fact, when the motions of the new planet and the earth combine to bring them to their positions of greatest proximity, the new planet will approach us closer than any other celestial body except our own moon. Witt named his new planet Eros. Its size, though small, may prove to be sufficient to bring it within the possibilities of naked-eye observation at the time of closest approach to the earth.

To astronomers the great importance of this new planet is due to the following circumstance: For certain reasons too technical to be stated here in detail, the distance from the earth to any planet can be determined with a degree of precision which is greatest for planets that are near us. Thus in time we shall learn the distance of Eros more accurately than we know any other celestial distance. From this, by a process of calculation, the solar distance from the earth is determinable. But the distance from earth to sun is the fundamental astronomical unit of measure; so that Witt's discovery, through its effect on the unit of measure, will doubtless influence every part of the science of astronomy. Here we have once more a striking instance of the reward sure to overtake the diligent worker in science—a whole generation of men will doubtless pass away before we shall have exhausted the scientific advantages to be drawn from Witt's remarkable observation of 1898.

HOW TO MAKE A SUN-DIAL[A]

Long before clocks and watches had been invented, people began to measure time with sun-dials. Nowadays, when almost everyone has a watch in his pocket, and can have a clock, too, on the mantel-piece of every room in the house, the sun-dial has ceased to be needed in ordinary life. But it is still just as interesting as ever to anyone who would like to have the means of getting time direct from the sun, the great hour-hand or timekeeper of the sky. Any person who is handy with tools can make a sun-dial quite easily, by following the directions given below.

In the first place, you must know that the sun-dial gives the time by means of the sun's shadow. If you stick a walking-cane up in the sand on a bright, sunshiny day, the cane has a long shadow that looks like a dark line on the ground. Now if you watch this shadow carefully, you will see that it does not stay in the same place all day. Slowly but surely, as the sun climbs up in the sky, the shadow creeps around the cane. You can see quite easily that if the cane were fastened in a board floor, and if we could mark on the floor the places where the shadow was at different hours of the day, we could make the shadow tell us the time just like the hour-hand of a clock. A sun-dial is just such an arrangement as this, and I will show you how to mark the shadow places exactly, so as to tell the right time without any trouble whenever the sun shines.

If you were to watch very carefully such an arrangement as a cane standing in a board floor, you would not find the creeping shadow in just the same place at the same time every day. If you marked the place of the shadow at exactly ten o'clock by your watch some morning, and then went back another day at ten, you would not find the shadow on the old mark. It would not get very far from it in a day or two, but in a month or so it would be quite a distance away. Now, of course, a sun-dial would be of no use if it did not tell the time correctly every day; and in fact, it is not easy to make a dial when the shadow is cast by a stick standing straight up. But we can get over this difficulty very well by letting the shadow be cast by a stick that leans over toward the floor just the right amount, as I will explain in a moment. Of course, we should not really use the floor for our sun-dial. It is much better to mark out the hour-lines, as they are called, on a smooth piece of ordinary white board, and then, after the dial is finished, it can be screwed down to a piazza floor or railing, or it can be fastened on a window-sill. It ought to be put in a place where the sun can get at it most of the time, because, of course, you cannot use the sun-dial when the sun is not shining on it. If the dial is set on a window-sill (of a city house, for instance) you must choose a south window if you can, so as to get the sun nearly all day. If you have to take an east window, you can use the dial in the morning only, and in a west window only in the afternoon. Sometimes it is best not to try to fasten the dial to its support with screws, but just to mark its place, and then set it out whenever you want to use it. For if the dial is made of wood, and not painted, it might be injured by rain or snow in bad weather if left out on a window-sill or piazza.

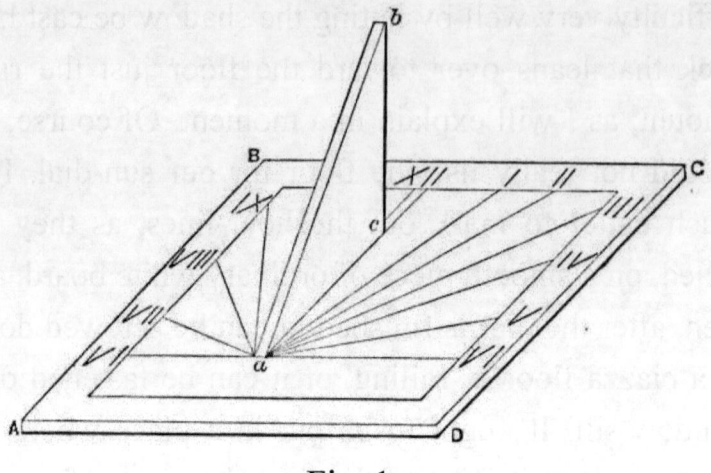

Fig. 1.

It is not quite easy to fasten a little stick to a board so that it will lean over just right. So it is better not to use a stick or a cane in the way I have described, but instead to use a piece of board cut to just the right shape.

Fig. 1 shows what a sun-dial should look like. The lines to show the shadow's place at the different hours of the day will be marked on the board ABCD, and this will be put flat on the window-sill or piazza floor. The three-cornered piece of board *abc* is fastened to the bottom-board ABCD by screws going through ABCD from underneath. The edge *ab* of the three-cornered board *abc* then takes the place of the leaning stick or cane, and the time is marked by the shadow cast by the edge *ab*. Of course, it is important that this edge should be straight and perfectly flat and even. If you are handy with tools, you can make it quite easily, but if not, you can mark the right shape on a piece of paper very carefully, and take it to a carpenter, who can cut the board according to the pattern you have marked on the paper.

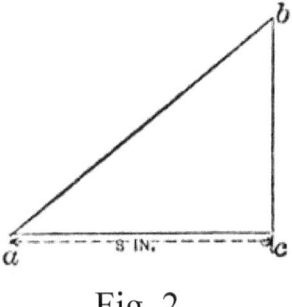

Fig. 2.

Now I must tell you how to draw the shape of the three-cornered board *abc*. Fig. 2 shows how it is done. The side *ac* should always be just five inches long. The side *bc* is drawn at right angles to *ac*, which you can do with an ordinary carpenter's square. The length of *bc* depends on the place for which the dial is made. The following table gives the length of *bc* for various places in the United States, and, after you have marked out the length of *bc*, it is only necessary to complete the three-cornered piece by drawing the side *ab* from *a* to *b*.

TABLE SHOWING THE LENGTH OF THE SIDE *bc*.

Place.	*bc* Inches.		Place.	*bc* Inches.	
Albany	4	11-16	New York	4	3-8
Baltimore	4	1-16	Omaha	4	3-8
Boston	4	1-2	Philadelphia	4	3-16
Buffalo	4	11-16	Pittsburg	4	3-8
Charleston	3	1-4	Portland, Me	4	13-16
Chicago	4	1-2	Richmond	3	15-16
Cincinnati	4	1-16	Rochester	4	11-16
Cleveland	4	1-2	San Diego	3	1-4
Denver	4	3-16	San Francisco	3	15-16
Detroit	4	1-2	Savannah	3	1-8
Indianapolis	4	1-16	St. Louis	3	15-16
Kansas City	3	15-16	St. Paul	5	
Louisville	3	15-16	Seattle	5	9-16
Milwaukee	3	11-16	Washington, D. C.	4	1-16
New Orleans	2	7-8			

If you wish to make a dial for a place not given in the table, it will be near enough to use the distance *bc* as given for the place nearest to you. But in selecting the nearest place from the table, please remember to take that one of the cities mentioned which is nearest to you in a north-and-south direction. It does not matter how far away the place is in an east-and-west direction. So, instead of taking the place that is nearest to you on the map in a straight line, take the place to which you could travel by going principally east or west, and very little north or south. The figure drawn is about the right shape for New York. The board used for the three-cornered piece should be about one-half inch thick. But if you are making a window-sill dial, you may prefer to have it smaller than I have described. You can easily have it half as big by making all the sizes and lines in half-inches where the table calls for inches.

After you have marked out the dimensions for the three-cornered piece that is to throw the shadow, you can prepare the dial itself, with the lines that mark the place of the shadow for every hour of the day. This you can do in the manner shown in Fig. 3. Just as in the case of the three-cornered piece, you can draw the dial with a pencil directly on a smooth piece of white board, about three-quarters of an inch thick, or you can mark it out on a paper pattern and transfer it afterward to the board. Perhaps it will be as well to begin by drawing on paper, as any mistakes can then be corrected before you commence to mark your wood.

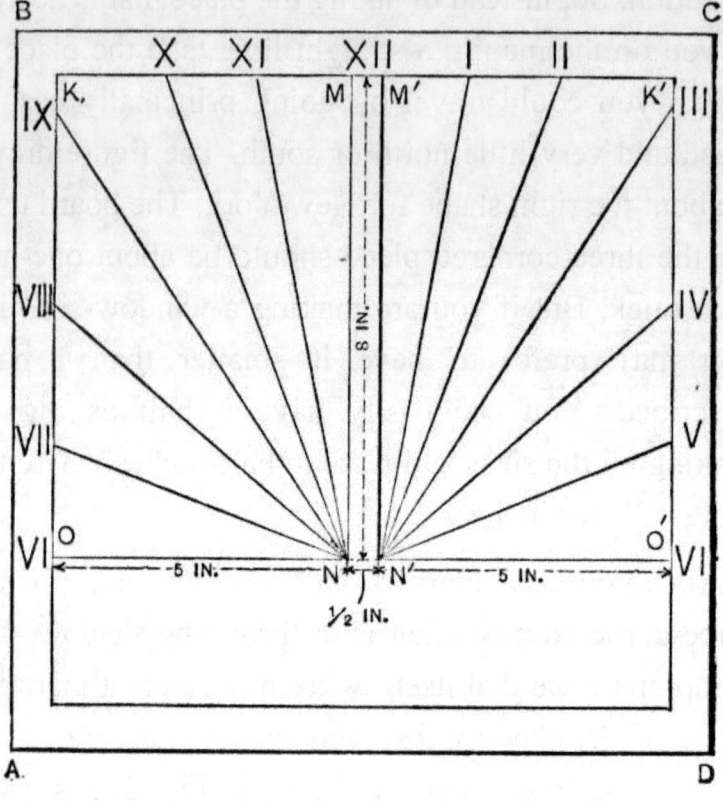

Fig. 3.

In the first place you must draw a couple of lines MN and M'N', eight inches long, and just far enough apart to fit the edge of your three-cornered shadow-piece. You will remember I told you to make that one-half inch thick, so your two lines will also be one-half inch apart. Now draw the two lines NO and N'O' square with MN and M'N', and make the distances NO and N'O' just five inches each. The lines OK, O'K', and the other lines forming the outer border of the dial, are then drawn just as shown, OK and O'K' being just eight inches long, the same as MN and M'N'. The lower lines in the figure, which are not very important, are to complete the squares. You must mark the lines NO and N'O' with the figures VI, these being the lines reached by the shadow at six o'clock in the morning and evening. The points where the VII, VIII, and other hour-lines cut the lines OK, O'K', MK, and M'K' can be found from the table on page 78.

In using the table you will notice that the line IX falls sometimes on one side of the corner K, and sometimes on the other. Thus for Albany the line passes seven and seven-sixteenth inches from O, while for Charleston it passes four and three-eighth inches from M. For Baltimore it passes exactly through the corner K.

TABLE SHOWING HOW TO MARK THE HOUR-LINES.

Place.	Distance from O to the line marked			Distance from M to the line marked		
	VII.	VIII.	IX.	IX.	X.	XI.
	Inches.	Inches.	Inches.	Inches.	Inches.	Inches.
Albany	1 15-16	4 3-16	7 7-16		3 1-16	1 7-16
Baltimore	2 1-8	4 11-16	8		2 7-8	1 7-16
Boston	2	4 5-16	7 7-16		3 1-16	1 7-16
Buffalo	1 15-16	4 3-16	7 7-16		3 1-16	1 7-16
Charleston	2 7-16	5 3-8		4 3-8	2 1-2	1 1-8

Place.	Distance from O to the line marked			Distance from M to the line marked		
	VII.	VIII.	IX.	IX.	X.	XI.
Chicago	2	4 5-16	7 7-16		3 1-16	1 7-16
Cincinnati	2 1-8	4 11-16	8		2 7-8	1 7-16
Cleveland	2	4 5-16	7 7-16	—	3 1-16	1 7-16
Denver	2 1-8	4 1-2	7 11-16		2 7-8	1 7-16
Detroit	2	4 5-16	7 7-16		3 1-16	1 7-16
Indianapolis	2 1-8	4 11-16	8		2 7-8	1 7-16
Kansas City	2 1-4	4 11-16	8		2 7-8	1 5-16
Louisville	2 1-4	4 11-16	8		2 7-8	1 5-16
Milwaukee	1 15-16	4 3-16	7 7-16		3 1-16	1 7-16
New Orleans	2 11-16	5 3-4		4 1-16	2 5-16	1 1-8
New York	2	4 5-16	7 11-16		3 1-16	1 7-16
Omaha	2	4 5-16	7 11-16		3 1-16	1 7-16
Philadelphia	2 1-8	4 1-2	7 11-16		2 7-8	1 7-16
Pittsburg	2	4 5-16	7 11-16		3 1-16	1 7-16
Portland, Me	1 15-16	4 3-16	7 1-8		3 3-16	1 1-2
Richmond	2 1-4	4 11-16	8		2 7-8	1 5-16
Rochester	1 15-16	4 3-16	7 7-16		3 1-16	1 7-16
San Diego	2 7-16	5 3-8		4 3-8	2 1-2	1 1-8
San Francisco	2 1-4	4 11-16	8		2 7-8	1 5-16
Savannah	2 9-16	5 9-16		4 1-4	2 1-2	1 1-8
St. Louis	2 1-4	4 11-16	8		2 7-8	1 5-16
St. Paul	1 15-16	4 1-16	7 1-8		3 3-16	1 1-2
Seattle	1 13-16	3 15-16	6 5-8		3 3-8	1 1-2
Washington, D. C.	2 1-8	4 11-16	8		2 7-8	1 7-16

The distance for the line marked V from O' is just the same as the distance from O to VII. Similarly, IV corresponds to VIII, III to IX, II to X, and I to XI. The number XII is marked at MM' as shown. If you desire to add lines (not shown in Fig. 3 to avoid confusion) for hours earlier than six in the morning, it is merely necessary to mark off a distance on the line KO, below the point O, and equal to the distance from O to VII. This will give the point where the 5 A.M. shadow line drawn from N cuts the line KO. A corresponding line for 7 P.M. can be drawn from N' on the other side of the figure.

After you have marked out the dial very carefully, you must fasten the three-cornered shadow-piece to it in such a way that the whole instrument will look like Fig. 1. The edge *ac* (Fig. 2) goes on NM (Fig. 3). The point *a* (Fig. 2) must come exactly on N (Fig. 3); and as the lines NM (Fig. 3) and N'M' (Fig. 3) have been made just the right distance apart to fit the thickness of the three-cornered piece *abc* (Fig. 2), everything will go together just right. The point *c* (Fig. 2) will not quite reach to M (Fig. 3), but will be on the line NM (Fig. 3) at a distance of three inches from M. The two pieces of wood will be fastened together with three screws going through the bottom-board ABCD (Figs. 1 and 3) and into the edge *ac* (Fig. 2) of the three-cornered piece. The whole instrument will then look something like Fig. 1.

After you have got your sun-dial put together, you need only set it in the sun in a level place, on a piazza or window-sill, and turn it round until it tells the right time by the shadow. You can get your local time from a watch near enough for setting up the dial. Once the dial is set right you can screw it down or mark its position, and it will continue to give correct solar time every day in the year.

If you wish to adjust the dial very closely, you must go out some fine day and note the error of the dial by a watch at about ten in the morning, and at noon, and again at about two in the afternoon. If the error is the same each time, the dial is rightly set. If not, you must try, by turning the dial slightly, to get it so placed that your three errors will be nearly the same. When you have got them as nearly alike as you can, the dial will be sufficiently near right. The solar or dial time may, however, differ somewhat from ordinary watch time, but the difference will never be great enough to matter, when we remember that sun-dials are only rough timekeepers after all, and useful principally for amusement.

FOOTNOTE:

[A]This chapter is especially intended for boys and girls and others who like to make things with carpenters' tools.

PHOTOGRAPHY IN ASTRONOMY

New highways of science have been monumented now and again by the masterful efforts of genius, working single-handed; but more often it is slow-moving time that ripens discovery, and, at the proper moment, opens some new path to men whose intellectual power is but willingness to learn. So the annals of astronomical photography do not recount the achievements of extraordinary genius. It would have been strange, indeed, if the discovery of photography had not been followed by its application to astronomy.

The whole range of chemical science contains no experiment of greater inherent interest than the development of a photographic plate. Let but the smallest ray of light fall upon its strangely sensitive surface, and some subtle invisible change takes place. It is then merely necessary to plunge the plate into a properly prepared chemical bath, and the gradual process of developing the picture begins. Slowly, very slowly, the colorless surface darkens wherever light has touched it. Let us imagine that the exposure has been made with an ordinary lens and camera, and that it is a landscape seeming to grow beneath the experimenter's eyes. At first only the most conspicuous objects make their appearance. But gradually the process extends, until finally every tiny detail is reproduced with marvellous fidelity to the original. The photographic plate, when developed in this way, is called a "negative." For in Nature luminous points, or sources of light, are bright, while the developing negative turns dark wherever light has acted. Thus the negative, while true to Nature, reproduces everything in a reversed way; bright things are dark, and shadows appear light. For ordinary purposes, therefore, the negative has to be replaced by a new photograph made by copying it again photographically. In this way it is again reversed, giving us a picture corresponding correctly to the facts as seen. Such a copy from a negative is what is ordinarily called a photograph; technically, it is known as a "positive."

One of the remarkable things about the sensitive plate is its complete indifference to the distance from which the light comes. It is ready to yield obediently to the ray of some distant star that may have journeyed, as it were, from the very vanishing point of space, or to the bright glow of an electric light upon the photographer's table. This quality makes its use especially advantageous in astronomy, since we can gain knowledge of remote stars only by a study of the light they send us. In such study the photographic plate possesses a supreme advantage over the human eye. If the conditions of weather and atmosphere are favorable, an observer looking through an ordinary telescope will see nearly as much at the first glance as he will ever see. Attentive and continued study will enable him to fix details upon his memory, and to record them by means of drawings and diagrams. Occasional moments of especially undisturbed atmospheric conditions will allow him to glimpse faint objects seldom visible. But on the whole, telescopic astronomers add little to their harvest by continued husbandry in the same field of stars. Photography is different. The effect of light upon the sensitive surface of the plate is strictly cumulative. If a given star can bring about a certain result when it has been allowed to act upon the plate for one minute, then in two or three minutes it will accomplish much more. Perhaps a single minute's exposure would have produced a mark scarcely perceptible upon the developed negative. In that case, three or four minutes would give us a perfectly well defined black image of the star.

Star-Field in Constellation Monoceros.
Photographed by Barnard, February 1, 1894.
Exposure, three hours.

Thus, by lengthening the exposure we can make the fainter stars impress themselves upon the plate. If their light is not able to produce the desired effect in minutes, we can let its action accumulate for hours. In this manner it becomes possible and easy to photograph objects so faint that they have never been seen, even with our most powerful telescopes. This achievement ranks high among those which make astronomy appeal so strongly to the imagination. Scientific men are not given to fancies; nor should they be. But the first long-exposure photograph must have been an exciting thing. After coming from the observatory, the chemical development was, of course, made in a dark room, so that no additional light might harm the plate until the process was complete. Carrying it out then into the light, that early experimenter cannot but have felt a thrill of triumph; for his hand held a true picture of dim stars to the eye unlighted, lifted into view as if by magic.

Plates have been thus exposed as long as twenty-five hours, and the manner of doing it is very interesting. Of course, it is impossible to carry on the work continuously for so long a period, since the beginning of daylight would surely ruin the photograph. In fact, the astronomer must stop before even the faintest streak of dawn begins to redden the eastern sky. Moreover, making astronomical negatives requires excessively close attention, and this it is impossible to give continuously during more than a few hours. But the exposure of a single plate can be extended over several nights without difficulty. It is merely necessary to close the plate-holder with a "light-tight" cover when the first night's work is finished. To begin further exposure of the same plate on another night, we simply aim the photographic telescope at precisely the same point of the sky as before. The light-tight plate-holder being again opened, the exposure can go on as if there had been no interruption.

Astronomers have invented a most ingenious device for making sure that the telescope's aim can be brought back again to the same point with great exactness. This is a very important matter; for the slightest disturbance of the plate before the second or subsequent portions of the exposure would ruin everything. Instead of a very complete single picture, we should have two partial ones mixed up together in inextricable confusion.

To prevent this, photographic telescopes are made double, not altogether unlike an opera-glass. One of the tubes is arranged for photography proper, while the other is fitted with lenses suitable for an ordinary visual telescope. The two tubes are made parallel. Thus the astronomer, by looking through the visual glass, can watch objects in the heavens even while they are being photographed. The visual half of the instrument is provided with a pair of very fine cross-wires movable at will in the field of view. These can be made to bisect some little star exactly, before beginning the first night's work. Afterward, everything about the instrument having been left unchanged, the astronomer can always assure himself of coming back to precisely the same point of the sky, by so adjusting the instrument that the same little star is again bisected.

It must not be supposed, however, that the entire instrument remains unmoved, even during the whole of a single night's exposure. For in that case, the apparent motion of the stars as they rise or set in the sky would speedily carry them out of the telescope's field of view. Consequently, this motion has to be counteracted by shifting the telescope so as to follow the stars. This can be accomplished accurately and automatically by means of clock-work mechanism. Such contrivances have already been applied in the past to visual telescopes, because even then they facilitated the observer's work. They save him the trouble of turning his instrument every few minutes, and allow him to give his undivided attention to the actual business of observation.

For photographic purposes the telescope needs to "follow" the stars far more accurately than in the older kind of observing with the eye. Nor is it possible to make a clock that will drive the instrument satisfactorily and quite automatically. But by means of the second or visual telescope, astronomers can always ascertain whether the clock is working correctly at any given moment. It requires only a glance at the little star bisected by the cross-wires, and, if there has been the slightest imperfection in the following by clock-work, the star will no longer be cut exactly by the wires.

The astronomer can at once correct any error by putting in operation a very ingenious mechanical device sometimes called a "mouse-control." He need only touch an electric button, and a signal is sent into the clock-work. Instantly there is a shifting of the mechanism. For one of the regular driving wheels is substituted, temporarily, another having an *extra tooth*. This makes the clock run a little faster so long as the electric current passes. In a similar way, by means of another button, the clock can be made to run slower temporarily. Thus by watching the cross-wires continuously, and manipulating his two electric buttons, the photographic astronomer can compel his telescope to follow exactly the object under observation, and he can make certain of obtaining a perfect negative.

These long-exposure plates are intended especially for what may be called descriptive astronomy. With them, as we have seen, advantage is taken of cumulative light-effects on the sensitive plate, and the telescope's light-gathering and space-penetrating powers are vastly increased. We are enabled to carry our researches far beyond the confines of the old visible universe. Extremely faint objects can be recorded, even down to their minutest details, with a fidelity unknown to older visual methods. But at present we intend to consider principally applications of photography in the astronomy of measurement, rather than the descriptive branch of our subject. Instead of describing pictures made simply to see what certain objects look like in the sky, we shall consider negatives intended for precise measurement, with all that the word precision implies in celestial science.

Taking up first the photography of stars, we must begin by mentioning the work of Rutherfurd at New York. More than thirty years ago he had so far perfected methods of stellar photography that he was able to secure excellent pictures of stars as faint as the ninth magnitude. In those days the modern process of dry-plate photography had not been invented. To-day, plates exposed in the photographic telescope are made of glass covered with a perfectly dry film of sensitized gelatine. But in the old wet-plate process the sensitive film was first wetted with a chemical solution; and this solution could not be allowed to dry during the exposure. Consequently, Rutherfurd was limited to exposures a few minutes in length, while nowadays, as we have said, their duration can be prolonged at will.

When we add to this the fact that the old plates were far less sensitive to light than those now available, it is easy to see what were the difficulties in the way of photographing faint stars in Rutherfurd's time. Nor did he possess the modern ingenious device of a combined visual and photographic instrument. He had no electric controlling apparatus. In fact, the younger generation of astronomers can form no adequate idea of the patience and personal skill Rutherfurd must have had at his command. For he certainly did produce negatives that are but little inferior to the best that can be made to-day. His only limitation was that he could not obtain images of stars much below the ninth magnitude.

To understand just what is meant here by the ninth magnitude, it is necessary to go back in imagination to the time of Hipparchus, the father of sidereal astronomy. (See page 39.) He adopted the convenient plan of dividing all the stars visible to the naked eye (of course, he had no telescope) into six classes, according to their brilliancy. The faintest visible stars were put in the sixth class, and all the others were assigned somewhat arbitrarily to one or the other of the brighter classes.

Modern astronomers have devised a more scientific system, which has been made to conform very nearly to that of Hipparchus, just as it has come down to us through the ages. We have adopted a certain arbitrary degree of luminosity as the standard "first-magnitude"; compared with sunlight, this may be represented roughly by a fraction of which the numerator is 1, and the denominator about eighty thousand millions. The standard second-magnitude star is one whose light, compared with a first-magnitude, may be represented approximately by the fraction ⅖. The third magnitude, in turn, may be compared with the second by the same fraction ⅖; and so the classification is extended to magnitudes below those visible to the unaided eye. Each magnitude compares with the one above it, as the light of two candles would compare with the light of five.

Rutherfurd did not stop with mere photographs. He realized very clearly the obvious truth that by making a picture of the sky we simply change the scene of our operations. Upon the photograph we can measure that which we might have studied directly in the heavens; but so long as they remain unmeasured, celestial pictures have a potential value only. Locked within them may lie hidden some secret of our universe. But it will not come forth unsought. Patient effort must precede discovery, in photography, as elsewhere in science. There is no royal road. Rutherfurd devised an elaborate measuring-machine in which his photographs could be examined under the microscope with the most minute exactness. With this machine he measured a large number of his pictures; and it has been shown quite recently that the results obtained from them are comparable in accuracy with those coming from the most highly accredited methods of direct eye-observation.

And photographs are far superior in ease of manipulation. Convenient day-observing under the microscope in a comfortable astronomical laboratory is substituted for all the discomforts of a midnight vigil under the stars. The work of measurement can proceed in all weathers, whereas formerly it was limited strictly to perfectly clear nights. Lastly, the negatives form a permanent record, to which we can always return to correct errors or re-examine doubtful points.

Rutherfurd's stellar work extended down to about 1877, and included especially parallax determinations and the photography of star-clusters. Each of these subjects is receiving close attention from later investigators, and, therefore, merits brief mention here. Stellar parallax is in one sense but another name for stellar distance. Its measurement has been one of the important problems of astronomy for centuries, ever since men recognized that the Copernican theory of our universe requires the determination of stellar distances for its complete demonstration.

If the earth is swinging around the sun once a year in a mighty path or orbit, there must be changes of its position in space comparable in size with the orbit itself. And the stars ought to shift their apparent places on the sky to correspond with these changes in the terrestrial observer's position. The phenomenon is analogous to what occurs when we look out of a room, first through one window, and then through another. Any object on the opposite side of the street will be seen in a changed direction, on account of the observer's having shifted his position from one window to the other. If the object seemed to be due north when seen from the first window, it will, perhaps, appear a little east of north from the other. But this change of direction will be comparatively small, if the object under observation is very far away, in comparison with the distance between the two windows.

This is what occurs with the stars. The earth's orbit, vast as it is, shrinks into almost absolute insignificance when compared with the profound distances by which we are sundered from even the nearest fixed stars. Consequently, the shifting of their positions is also very small—so small as to be near the extreme limit separating that which is measurable from that which is beyond human ken.

Photography lends itself most readily to a study of this matter. Suppose a certain star is suspected of "having a parallax." In other words, we have reason to believe it near enough to admit of a successful measurement of distance. Perhaps it is a very bright star; and, other things being equal, it is probably fair to assume that brightness signifies nearness. And astronomers have certain other indications of proximity that guide them in the selection of proper objects for investigation, though such evidence, of course, never takes the place of actual measurement.

The star under examination is sure to have near it on the sky a number of stars so very small that we may safely take them to be immeasurably far away. The parallax star is among them, but not of them. We see it projected upon the background of the heavens, though it may in reality be quite near us, astronomically speaking. If this is really so, and the star, therefore, subject to the slight parallactic shifting already mentioned, we can detect it by noting the suspected star's position among the surrounding small stars. For these, being immeasurably remote, will remain unchanged, within the limits of our powers of observation, and thus serve as points of reference for marking the apparent shifting of the brighter star we are actually considering.

We have merely to photograph the region at various seasons of the year. Careful examination of the photographs under the microscope will then enable us to measure the slightest displacement of the parallax star. From these measures, by a process of calculation, astronomers can then obtain the star's distance. It will not become known in miles; we shall only ascertain how many times the distance between the earth and sun would have to be laid down like a measuring-rod, in order to cover the space separating us from the star: and the subsequent evaluation of this distance "earth to sun" in miles is another important problem in whose solution photography promises to be most useful.

The above method of measuring stellar distance is, of course, subject to whatever slight uncertainty arises from the assumption that the small stars used for comparison are themselves beyond the possibility of parallactic shifting. But astronomy possesses no better method. Moreover, the number of small stars used in this way is, of course, much larger in photography than it ever can be in visual work. In the former process, all surrounding stars can be photographed at once; in the latter each star must be measured separately, and daylight soon intervenes to impose a limit on numbers. Usually only two can be used; so that here photography has a most important advantage. It minimizes the chance of our parallax being rendered erroneous, by the stars of comparison not being really infinitely remote. This might happen, perhaps, in the case of one or two; but with an average result from a large number we know it to be practically impossible.

Cluster work is not altogether unlike "parallax hunting" in its preliminary stage of securing the photographic observations. The object is to obtain an absolutely faithful picture of a star group, just as it exists in the sky. We have every reason to suppose that a very large number of stars condensed into one small spot upon the heavens means something more than chance aggregation. The Pleiades group (page 10) contains thousands of massive stars, doubtless held together by the force of their mutual gravitational attraction. If this be true, there must be complex orbital motion in the cluster; and, as time goes on, we should actually see the separate components change their relative positions, as it were, before our eyes.

The details of such motion upon the great scale of cosmic space offer one of the many problems that make astronomy the grandest of human sciences.

We have said that time must pass before we can see these things; there may be centuries of waiting. But one way exists to hurry on the perfection of our knowledge; we must increase the precision of observations. Motions that would need the growth of centuries to become visible to the older astronomical appliances, might yield in a few decades to more delicate observational processes. Here photography is most promising. Having once obtained a surpassingly accurate picture of a star-cluster, we can subject it easily to precise microscopic measurement. The same operations repeated at a later date will enable us to compare the two series of measures, and thus ascertain the motions that may have occurred in the interval. The Rutherfurd photographs furnish a veritable mine of information in researches of this kind; for they antedate all other celestial photographs of precision by at least a quarter-century, and bring just so much nearer the time when definite knowledge shall replace information based on reasoning from probabilities.

Rutherfurd's methods showed the advantages of photography as applied to individual star-clusters. It required only the attention of some astronomer disposing of large observational facilities, and accustomed to operations upon a great scale, to apply similar methods throughout the whole heavens. In the year 1882 a bright comet was very conspicuous in the southern heavens. It was extensively observed from the southern hemisphere, and especially at the British Royal Observatory at the Cape of Good Hope.

Gill, director of that institution, conceived the idea that this comet might be bright enough to photograph. At that time, comet photography had been attempted but little, if at all, and it was by no means sure that the experiment would be successful. Nor was Gill well acquainted with the work of Rutherfurd; for the best results of that astronomer had lain dormant many years. He was one of those men with whom personal modesty amounts to a fault. Loath to put himself forward in any way, and disliking to rush into print, Rutherfurd had given but little publicity to his work. This peculiarity has, doubtless, delayed his just reputation; but he will lose nothing in the end from a brief postponement. Gill must, however, be credited with more penetration than would be his due if Rutherfurd had made it possible for others to know that he had anticipated many of the newer ideas.

However this may be, the comet was photographed with the help of a local portrait photographer named Allis. When Gill and Allis fastened a simple portrait camera belonging to the latter upon the tube of one of the Cape telescopes, and pointed it at the great comet, they little thought the experiment would lead to one of the greatest astronomical works ever attempted by men. Yet this was destined to occur. The negative they obtained showed an excellent picture of the comet; but what was more important for the future of sidereal astronomy, it was also quite thickly dotted with little black points corresponding to stars. The extraordinary ease with which the whole heavens could be thus charted photographically was brought home to Gill as never before. It was this comet picture that interested him in the application of photography to star-charting; and without his interest the now famous astro-photographic catalogue of the heavens would probably never have been made.

After considerable preliminary correspondence, a congress of astronomers was finally called to meet at Paris in 1887. Representatives of the principal observatories and civilized governments were present. They decided that the end of the nineteenth century should see the making of a great catalogue of all the stars in the sky, upon a scale of completeness and precision surpassing anything previously attempted. It is impossible to exaggerate the importance of such a work; for upon our star-catalogues depends ultimately the entire structure of astronomical science.

The work was far too vast for the powers of any observatory alone. Therefore, the whole sky, from pole to pole, was divided into eighteen belts or zones of approximately equal area; and each of these was assigned to a single observatory to be photographed. A series of telescopes was specially constructed, so that every part of the work should be done with the same type of instrument. As far as possible, an attempt was made to secure uniformity of methods, and particularly a uniform scale of precision. To cover the entire sky upon the plan proposed no less than 44,108 negatives are required; and most of these have now been finished. The further measurement of the pictures and the drawing up of a vast printed star-catalogue are also well under way. One of the participating observatories, that at Potsdam, Germany, has published the first volume of its part of the catalogue. It is estimated that this observatory alone will require twenty quarto volumes to contain merely the final results of its work on the catalogue. Altogether not less than two million stars will find a place in this, our latest directory of the heavens.

Such wholesale methods of attacking problems of observational astronomy are particularly characteristic of photography. The great catalogue is, perhaps, the best illustration of this tendency; but of scarcely smaller interest, though less important in reality, is the photographic method of dealing with minor planets. We have already said (page 63) that in the space between the orbits of Mars and Jupiter several hundred small bodies are moving around the sun in ordinary planetary orbits. These bodies are called asteroids, or minor planets. The visual method of discovering unknown members of this group was painfully tedious; but photography has changed matters completely, and has added immensely to our knowledge of the asteroids.

Wolf, of Heidelberg, first made use of the new process for minor-planet discovery. His method is sufficiently ingenious to deserve brief mention again. A photograph of a suitable region of the sky was made with an exposure lasting two or three hours. Throughout all this time the instrument was manipulated so as to follow the motion of the heavens in the way we have already explained, so that each star would appear on the negative as a small, round, black dot.

But if a minor planet happened to be in the region covered by the plate, its photographic image would be very different. For the orbital motion of the planet about the sun would make it move a little among the stars even in the two or three hours during which the plate was exposed. This motion would be faithfully reproduced in the picture, so that the planet would appear as a short curved line rather than a well-defined dot like a star. Thus the presence of such a line-image infallibly denotes an asteroid.

Subsequent calculations are necessary to ascertain whether the object is a planet already known or a genuine new discovery. Wolf, and others using his method in recent years, have made immense additions to our catalogue of asteroids. Indeed, the matter was beginning to lose interest on account of the frequency and sameness of these discoveries, when the astronomical world was startled by the finding of the Planet of 1898. (Page 58.)

On August 27, 1898, Witt, of Berlin, discovered the small body that bears the number "433" in the list of minor planets, and has received the name Eros. Its important peculiarity consists in the exceptional position of the orbit. While all the other asteroids are farther from the sun than Mars, and less distant than Jupiter, Eros can pass within the orbit of the former. At times, therefore, it will approach our earth more closely than any other permanent member of the solar system, excepting our own moon. So it is, in a sense, our nearest neighbor; and this fact alone makes it the most interesting of all the minor planets. The nineteenth century was opened by Piazzi's well-known discovery of the first of these bodies (page 59); it is, therefore, fitting that we should find the most important one at its close. We are almost certain that it will be possible to make use of Eros to solve with unprecedented accuracy the most important problem in all astronomy. This is the determination of our earth's distance from the sun. When considering stellar parallax, we have seen how our observations enable us to measure some of the stars' distances in terms of the distance "earth to sun" as a unit. It is, indeed, the fundamental unit for all astronomical measures, and its exact evaluation has always been considered the basal problem of astronomy. Astronomers know it as the problem of Solar Parallax.

We shall not here enter into the somewhat intricate details of this subject, however interesting they may be. The problem offers difficulties somewhat analogous to those confronting a surveyor who has to determine the distance of some inaccessible terrestrial point. To do this, it is necessary first to measure a "base-line," as we call it. Then the measurement of angles with a theodolite will make it possible to deduce the required distance of the inaccessible point by a process of calculation. To insure accuracy, however, as every surveyor knows, the base-line must be made long enough; and this is precisely what is impossible in the case of the solar parallax.

For we are necessarily limited to marking out our base-line on the earth; and the entire planet is too small to furnish one of really sufficient size. The best we can do is to use the distance between two observatories situated, as near as may be, on opposite sides of the earth. But even this base is wofully small. However, the smallness loses some of its harmful effect if we operate upon a planet that is comparatively near us. We can measure such a planet's distance more accurately than any other; and this being known, the solar distance can be computed by the aid of mathematical considerations based upon Newton's law of gravitation and observational determinations of the planetary orbital elements.

Photography is by no means limited to investigations in the older departments of astronomical observation. Its powerful arm has been stretched out to grasp as well the newer instruments of spectroscopic study. Here the sensitive plate has been substituted for the human eye with even greater relative advantage. The accurate microscopic measurement of difficult lines in stellar spectra was indeed possible by older methods; but photography has made it comparatively easy; and, above all, has rendered practicable series of observations extensive enough in numbers to furnish statistical information of real value. Only in this way have we been able to determine whether the stars, in their varied and unknown orbits, are approaching us or moving farther away. Even the speed of this approach or recession has become measurable, and has been evaluated in the case of many individual stars. (See page 21.)

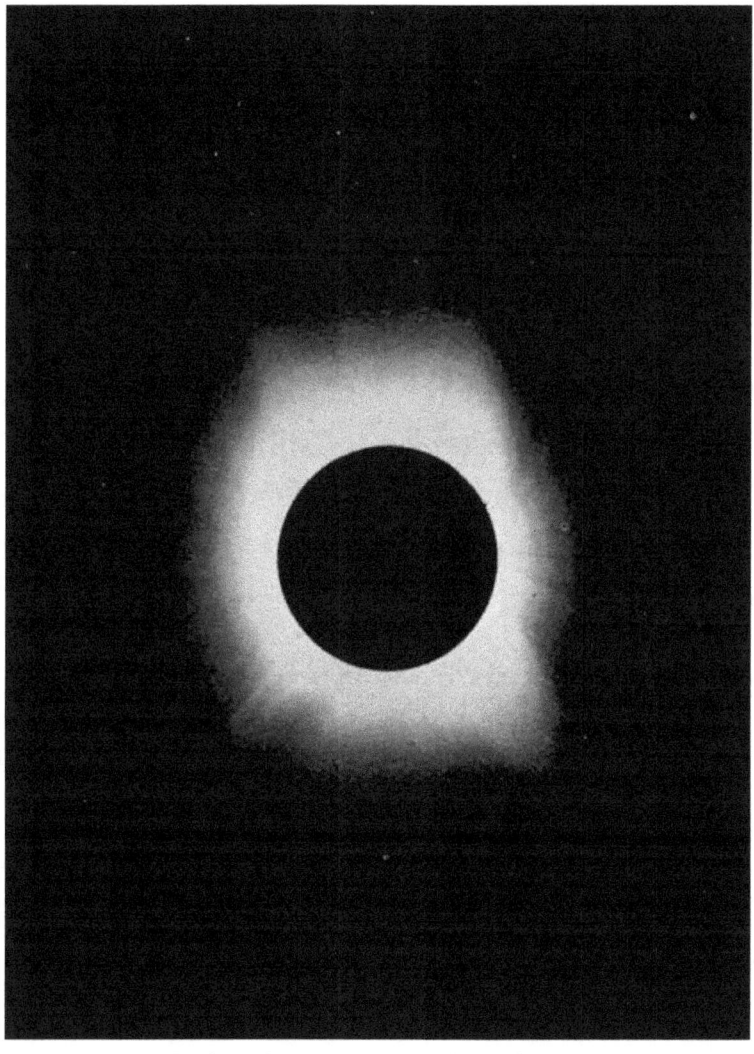

Solar Corona. Total Eclipse.
Photographed by Campbell, January 22, 1898; Jeur, India.

The subject of solar physics has become a veritable department of astronomy in the hands of photographic investigators. Ingenious spectro-photographic methods have been devised, whereby we have secured pictures of the sun from which we have learned much that must have remained forever unknown to older methods.

Especially useful has photography proved itself in the observation of total solar eclipses. It is only when the sun's bright disk is completely obscured by the interposed moon that we can see the faintly luminous structure of the solar corona, that great appendage of our sun, whose exact nature is still unexplained. Only during the few minutes of total eclipse in each century can we look upon it; and keen is the interest of astronomers when those few minutes occur. But it is found that eye observations made in hurried excitement have comparatively little value. Half a dozen persons might make drawings of the corona during the same eclipse, yet they would differ so much from one another as to leave the true outline very much in doubt. But with photography we can obtain a really correct picture whose details can be studied and discussed subsequently at leisure.

If we were asked to sum up in one word what photography has accomplished, we should say that observational astronomy has been revolutionized.

There is to-day scarcely an instrument of precision in which the sensitive plate has not been substituted for the human eye; scarcely an inquiry possible to the older method which cannot now be undertaken upon a grander scale. Novel investigations formerly not even possible are now entirely practicable by photography; and the end is not yet. Valuable as are the achievements already consummated, photography is richest in its promise for the future. Astronomy has been called the "perfect science"; it is safe to predict that the next generation will wonder that the knowledge we have to-day should ever have received so proud a title.

TIME STANDARDS OF THE WORLD

The question is often asked, "What is the practical use of astronomy?" We know, of course, that men would profit greatly from a study of that science, even if it could not be turned to any immediate bread-and-butter use; for astronomy is essentially the science of big things, and it makes men bigger to fix their minds on problems that deal with vast distances and seemingly endless periods of time. No one can look upon the quietly shining stars without being impressed by the thought of how they burned—then as now—before he himself was born, and so shall continue after he has passed away—aye, even after his latest descendants shall have vanished from the earth. Of all the sciences, astronomy is at once the most beautiful poetically, and yet the one offering the grandest and most difficult problems to the intellect. A study of these problems has ever been a labor of love to the greatest minds; their solution has been counted justly among man's loftiest achievements.

And yet of all the difficult and abstruse sciences, astronomy is, perhaps, the one that comes into the ordinary practical daily life of the people more definitely and frequently than any other. There exist at least three things we owe to astronomy that must be regarded as quite indispensable, from a purely practical point of view. In the first place, let us consider the maps in a work on geography. How many people ever think to ask how these maps are made? It is true that the ordinary processes of the surveyor would enable us to draw a map showing the outlines of a part of the earth's surface. Even the locations of towns and rivers might be marked in this way. But one of the most important things of all could not be added without the aid of astronomical observations. The latitude and longitude lines, which are essential to show the relation of the map to the rest of the earth, we owe to astronomy. The longitude lines, particularly, as we shall see farther on, play a most important part in the subject of time.

The second indispensable application of astronomy to ordinary business affairs relates to the subject of navigation. How do ships find their way across the ocean? There are no permanent marks on the sea, as there are on the land, by which the navigator can guide his course. Nevertheless, seamen know their path over the trackless ocean with a certainty as unerring as would be possible on shore; and it is all done by the help of astronomy. The navigator's observations of the sun are astronomical observations; the tables he uses in calculating his observations—the tables that tell him just where he is and in what direction he must go—are astronomical tables. Indeed, it is not too much to say that without astronomy there could be no safe ocean navigation.

But the third application of astronomy is of still greater importance in our daily life—the furnishing of correct time standards for all sorts of purposes. It is to this practical use of astronomical science that we would direct particular attention. Few persons ever think of the complicated machinery that must be put in motion in order to set a clock. A man forgets some evening to wind his watch at the accustomed hour. The next morning he finds it run down. It must be re-set. Most people simply go to the nearest clock, or ask some friend for the time, so as to start the watch correctly. More careful persons, perhaps, visit the jeweller's and take the time from his "regulator." But the regulator itself needs to be regulated. After all, it is nothing more than any other clock, except that greater care has been taken in the mechanical construction and arrangement of its various parts. Yet it is but a machine built by human hands, and, like all human works, it is necessarily imperfect. No matter how well it has been constructed, it will not run with perfectly rigid accuracy. Every day there will be a variation from the true time by a small amount, and in the course of days or weeks the accumulation of these successive small amounts will lead to a total of quite appreciable size.

Just as the ordinary citizen looks to the jeweller's regulator to correct his watch, so the jeweller applies to the astronomer for the correction of his regulator. Ever since the dawn of astronomy, in the earliest ages of which we have any record, the principal duty of the astronomer has been the furnishing of accurate time to the people. We shall not here enter into a detailed account, however interesting it would be, of the gradual development by which the very perfect system at present in use has been reached; but shall content ourselves with a description of the methods now employed in nearly all the civilized countries of the world.

In the first place, every observatory is, of course, provided with what is known as an astronomical clock. This instrument, from the astronomer's point of view, is something very different from the ordinary popular idea. To the average person an astronomical clock is a complicated and elaborate affair, giving the date, day of the week, phases of the moon, and other miscellaneous information. But in reality the astronomer wants none of these things. His one and only requirement is that the clock shall keep as near uniform time as may be possible to a machine constructed by human hands. No expense is spared in making the standard clock for an observatory. Real artists in mechanical construction—men who have attained a world-wide celebrity for delicate skill in fashioning the parts of a clock—such are the astronomer's clock-makers.

To increase precision of motion in the train of wheels, it is necessary that the mechanism be as simple as possible. For this reason all complications of date, etc., are left out. We have even abandoned the usual convenient plan of having the hour and minute hands mounted at the same centre; for this kind of mounting makes necessary a slightly more intricate form of wheelwork. The astronomer's clock usually has the centres of the second hand, minute hand, and hour hand in a straight line, and equally distant from each other. Each hand has its own dial; all drawn, of course, upon the same clock-face.

Even after such a clock has been made as accurately as possible, it will, nevertheless, not give the very best performance unless it is taken care of properly. It is necessary to mount it very firmly indeed. It should not be fastened to an ordinary wall, but a strong pier of masonry or brick must be built for it on a very solid foundation. Moreover, this pier is best placed underground in a cellar, so that the temperature of the clock can be kept nearly uniform all the year round; for we find that clocks do not run quite the same in hot weather as they do in cold. Makers have, indeed, tried to guard against this effect of temperature, by ingenious mechanical contrivances. But these are never quite perfect in their action, and it is best not to test them too severely by exposing the clock to sharp changes of heat and cold.

Another thing affecting the going of fine clocks, strange as it may seem, is the variation of barometric pressure. There is a slight but noticeable difference in their running when the barometer is high and when it is low. To prevent this, some of our best clocks have been enclosed in air-tight cases, so that outside barometric changes may not be felt in the least by the clock itself.

But even after all this has been accomplished, and the astronomer is in possession of a clock that may be called a masterpiece of mechanical construction, he is not any better off than was the jeweller with his regulator. After all, even the astronomical clock needs to be set, and its error must be determined from time to time. A final appeal must then be had to astronomical observations. The clock must be set by the stars and sun. For this purpose the astronomer uses an instrument called a "transit." This is simply a telescope of moderate size, possibly five or six feet long, and firmly attached to an axis at right angles to the tube of the telescope.

This axis is supported horizontally in such a way that it points as nearly as may be exactly east and west. The telescope itself being square with the axis, always points in a north-and-south direction. It is possible to rotate the telescope about its axis so as to reach all parts of the sky that are directly north or south of the observatory. In the field of view of the telescope certain very fine threads are mounted so as to form a little cross. As the telescope is rotated this cross traces out, as it were, a great circle on the sky; and this great circle is called the astronomical meridian.

Now we are in possession of certain star-tables, computed from the combined observations of astronomers in the last 150 years. These tables tell us the exact moment of time when any star is on the meridian. To discover, therefore, whether our clock is right on any given night, it is merely necessary to watch a star with the telescope, and note the exact instant by the clock when it reaches the little cross in the field of view. Knowing from the astronomical tables the time when the star ought to have been on the meridian, and having observed the clock time when it is actually there, the difference is, of course, the error of the clock. The result can be checked by observations of other stars, and the slight personal errors of observation can be rendered harmless by taking the mean from several stars. By an hour's work on a fine night it is possible to fix the clock error quite easily within the one-twentieth part of a second.

We have not space to enter into the interesting details of the methods by which the astronomical transit is accurately set in the right position, and how any slight residual error in its setting can be eliminated from our results by certain processes of computation. It must suffice to say that practically all time determinations in the observatory depend substantially upon the procedure outlined above.

The observatory clock having been once set right by observations of the sky, its error can be re-determined every few days quite easily. Thus even the small irregularities of its nearly perfect mechanism can be prevented from accumulating until they might reach a harmful magnitude. But we obtain in this way only a correct standard of time within the observatory itself. How can this be made available for the general public? The problem is quite simple with the aid of the electric telegraph. We shall give a brief account of the methods now in use in New York City, and these may be taken as essentially representative of those employed elsewhere.

Every day, at noon precisely, an electric signal is sent out by the United States Naval Observatory in Washington. The signal is regulated by the standard clock of the observatory, of course taking account of star observations made on the next preceding fine night. This signal is received in the central New York office of the telegraph company, where it is used to keep correct a very fine clock, which may be called the time standard of the telegraph company. This clock, in turn, has automatic electric connections, by means of which it is made to send out signals over what are called "time wires" that go all over the city. Jewellers, and others who desire correct time, can arrange to have a small electric sounder in their offices connected with the time wires. Thus the ticks of the telegraph company's standard clock are repeated automatically in the jeweller's shop, and used for controlling the exactness of his regulator. This, in brief, is the method by which the astronomer's careful determination of correct time is transferred and distributed to the people at large.

Having thus outlined the manner of obtaining and distributing correct time, we shall now consider the question of time differences between different places on the earth. This is a matter which many persons find most perplexing, and yet it is essentially quite simple in principle. Travellers, of course, are well acquainted with the fact that their watches often need to be reset when they arrive at their destination. Yet few ever stop to ask the cause.

Let us consider for a moment our method of measuring time. We go by the sun. If we leave out of account some small irregularities of the sun's motion that are of no consequence for our present purpose, we may lay down this fundamental principle: When the sun reaches its highest position in the sky it is twelve o'clock or noon.

The sun, as everyone knows, rises each morning in the east, slowly goes up higher and higher in the sky, and at last begins to descend again toward the west. But it is clear that as the sun travels from east to west, it must pass over the eastern one of any two cities sooner than the western one. When it reaches its greatest height over a western city it has, therefore, already passed its greatest height over an eastern one. In other words, when it is noon, or twelve o'clock, in the western city, it is already after noon in the eastern city. This is the simple and evident cause of time differences in different parts of the country. Of any two places the eastern one always has later time than the western. When we consider the matter in this way there is not the slightest difficulty in understanding how time differences arise. They will, of course, be greatest for places that are very far apart in an east-and-west direction. And this brings us again to the subject of longitude, which, as we have already said, plays an important part in all questions relating to time; for longitude is used to measure the distance in an east-and-west direction between different parts of the earth.

If we consider the earth as a large ball we can imagine a series of great circles drawn on its surface and passing directly from the North Pole to the South Pole. Such a circle could be drawn through any point on the earth. If we imagine a pair of them drawn through two cities, such as New York and London, the longitude difference of these two cities is defined as the angle at the North Pole between the two great circles in question. The size of this angle can be expressed in degrees. If we then wish to know the difference in time between New York and London in hours, we need only divide their longitude difference in degrees by the number 15. In this simple way we can get the time difference of any two places. We merely measure the longitude difference on a map, and then divide by 15 to get the time difference. These time differences can sometimes become quite large. Indeed, for two places differing 180 degrees in longitude, the time difference will evidently be no less than twelve hours.

Most civilized nations have agreed informally to adopt some one city as the fundamental point from which all longitudes are to be counted. Up to the present we have considered only longitude differences; but when we speak of the longitude of a city we mean its longitude difference from the place chosen by common consent as the origin for measuring longitudes. The town almost universally used for this purpose is Greenwich, near London, England. Here is situated the British Royal Observatory, one of the oldest and most important institutions of its kind in the world. The great longitude circle passing through the centre of the astronomical transit at the Greenwich observatory is the fundamental longitude circle of the earth. The longitude of any other town is then simply the angle at the pole between the longitude circle through that town and the fundamental Greenwich one here described.

Longitudes are counted both eastward and westward from Greenwich. Thus New York is in 74 degrees west longitude, while Berlin is in 14 degrees east longitude. This has led to a rather curious state of affairs in those parts of the earth the longitudes of which are nearly 180 degrees east or west. There are a number of islands in that part of the world, and if we imagine for a moment one whose longitude is just 180 degrees, we shall have the following remarkable result as to its time difference from Greenwich.

We have seen that of any two places the eastern always has the later time. Now, since our imaginary island is exactly 180 degrees from Greenwich, we can consider it as being either 180 degrees east or 180 degrees west. But if we call it 180 degrees east, its time will be twelve hours later than Greenwich, and if we call it 180 degrees west, its time will be twelve hours earlier than Greenwich. Evidently there will be a difference of just twenty-four hours, or one whole day, between these two possible ways of reckoning its time. This circumstance has actually led to considerable confusion in some of the islands of the Pacific Ocean. The navigators who discovered the various islands naturally gave them the date which they brought from Europe. And as some of these navigators sailed eastward, around the Cape of Good Hope, and others westward, around Cape Horn, the dates they gave to the several islands differed by just one day.

The state of affairs at the present time has been adjusted by a sort of informal agreement. An arbitrary line has been drawn on the map near the 180th longitude circle, and it has been decided that the islands on the east side of this line shall count their longitudes west from Greenwich, and those west of the line shall count longitude east from Greenwich. Thus Samoa is nearly 180 degrees west of Greenwich, while the Fiji Islands are nearly 180 degrees east. Yet the islands are very near each other, though the arbitrary line passes between them. As a result, when it is Sunday in Samoa it is Monday in the Fiji Islands. The arbitrary line described here is sometimes called the International Date-Line.

It does not pass very near the Philippine Islands, which are situated in about 120 degrees east longitude, and, therefore, use a time about eight hours later than Greenwich. New York, being about 74 degrees west of Greenwich, is about five hours earlier in time. Consequently, as we may remark in passing, Philippine time is about thirteen hours later than New York time. Thus, five o'clock, Sunday morning, May 1st, in Manila, would correspond to four o'clock, Saturday afternoon, April 30th, in New York.

There is another kind of time which we shall explain briefly—the so-called "standard," or railroad time, which came into general use in the United States some few years ago, and has since been generally adopted throughout the world. It requires but a few moments' consideration to see that the accidental situation of the different large cities in any country will cause their local times to differ by odd numbers of hours, minutes, and seconds. Thus a great deal of inconvenience has been caused in the past. For instance, a train might leave New York at a certain hour by New York time. It would then arrive in Buffalo some hours later by New York time. But it would leave Buffalo by Buffalo time, which is quite different. Thus there would be a sort of jump in the time-table at Buffalo, and it would be a jump of an odd number of minutes.

It would be different in different cities, and very hard to remember. Indeed, as each railway usually ran its trains by the time used in the principal city along its line, it might happen that three or four different railroad times would be used in a single city where several roads met. This has all been avoided by introducing the standard time system. According to this the whole country is divided into a series of time zones, fifteen degrees wide, and so arranged that the middle line of each zone falls at a point whose longitude from Greenwich is 60, 75, 90, 105, or 120 degrees. The times at these middle lines are, therefore, earlier than Greenwich time by an even number of hours. Thus, for instance, the 75-degree line is just five even hours earlier than Greenwich time. All cities simply use the time of the nearest one of these special lines.

This does not result in doing away with time differences altogether—that would, of course, be impossible in the nature of things—but for the complicated odd differences in hours and minutes, we have substituted the infinitely simpler series of differences in even hours. The traveller from Chicago to New York can reset his watch by putting it just one hour later on his arrival—the minute hand is kept unchanged, and no New York timepiece need be consulted to set the watch right on arriving. There can be no doubt that this standard-time system must be considered one of the most important contributions of astronomical science to the convenience of man.

Its value has received the widest recognition, and its use has now extended to almost all civilized countries—France is the only nation of importance still remaining outside the time-zone system. In the following table we give the standard time of the various parts of the earth as compared with Greenwich, together with the date of adopting the new time system. It will be noticed that in certain cases even half-hours have been employed to separate the time-zones, instead of even hours as used in the United States.

TABLE OF THE WORLD'S TIME STANDARDS

When it is Noon at Greenwich it is	In	Date of Adopting Standard Time System.
Noon	Great Britain.	
	Belgium.	May, 1892.
	Holland.	May, 1892.
	Spain.	January, 1901.
1 P.M.	Germany.	April, 1893.
	Italy.	November, 1893.
	Denmark.	January, 1894.
	Switzerland.	June, 1894.
	Norway.	January, 1895.
	Austria (railways).	
1.30 P.M.	Cape Colony.	1892.
	Orange River Colony.	1892.
	Transvaal.	1892.
2 P.M.	Natal.	September, 1895.
	Turkey	

	(railways).	
	Egypt.	October, 1900.
8 P.M.	West Australia.	February, 1895.
9 P.M.	Japan.	1896.
9.30 P.M.	South Australia.	May, 1899.
10 P.M.	Victoria.	February, 1895.
	New South Wales.	February, 1895.
	Queensland.	February, 1895.
11 P.M.	New Zealand.	

In the United States and Canada it is

4 A.M. by	Pacific Time	when	it is	Noon	at	Greenwich.
5 A.M. "	Mountain	"	"	"	"	"
6 A.M. "	Central	"	"	"	"	"
7 A.M	Eastern	"	"	"	"	"

."
8
A.M   Colonial "   "   "   "   "
."

MOTIONS OF THE EARTH'S POLE

Students of geology have been puzzled for many years by traces remaining from the period when a large part of the earth was covered with a heavy cap of ice. These shreds of evidence all seem to point to the conclusion that the centre of the ice-covered region was quite far away from the present position of the north pole of the earth. If we are to regard the pole as very near the point of greatest cold, it becomes a matter of much interest to examine whether the pole has always occupied its present position, or whether it has been subject to slow changes of place upon the earth's surface. Therefore, the geologists have appealed to astronomers to discover whether they are in possession of any observational evidence tending to show that the pole is in motion.

Now we may say at once that astronomical research has not as yet revealed the evidence thus expected. Astronomy has been unable to come to the rescue of geological theory. From about the year 1750, which saw the beginning of precise observation in the modern sense, down to very recent times, astronomers were compelled to deny the possibility of any appreciable motion of the pole. Observational processes, it is true, furnished slightly divergent pole positions from time to time. Yet these discrepancies were always so minute as to be indistinguishable from those slight personal errors that are ever inseparable from results obtained by the fallible human eye.

But in the last few years improved methods of observation, coupled with extreme diligence in their application by astronomers generally, have brought to light a certain small motion of the pole which had never before been demonstrated in a reliable way. This motion, it is true, is not of the character demanded by geological theory, for the geologists had been led to expect a motion which would be continuous in the same direction, no matter how slow might be its annual amount; for the vast extent of geologic time would give even the slowest of motions an opportunity to produce large effects, provided its results could be continuously cumulative. Given time enough, and the pole might move anywhere on the earth, no matter how slow might be its tortoise speed.

But the small motion we have discovered is neither cumulative nor continuous in one direction. It is what we call a periodic motion, the pole swinging now to one side, and now to the other, of its mean or average position. Thus this new discovery cannot be said to unravel the mysterious puzzle of the geologists. Yet it is not without the keenest interest, even from their point of view; for the proof of any form of motion in a pole previously supposed to be absolutely at rest may mean everything. No man can say what results will be revealed by the further observations now being continued with great diligence.

In the first place, it is important to explain that any such motions as we have under consideration will show themselves to ordinary observational processes principally in the form of changes of terrestrial latitudes. Let us imagine a pair of straight lines passing through the centre of the earth and terminating, one at the observer's station on the earth's surface, and the other at that point of the equator which is nearest the observer. Then, according to the ordinary definition of latitude, the angle between these two imaginary lines is called the latitude of the point of observation. Now we know, of course, that the equator is everywhere just 90 degrees from the pole. Consequently, if the pole is subject to any motion at all, the equator must also partake of the motion.

Thus the angle between our two imaginary lines will be affected directly by polar movement, and the latitude obtained by astronomical observation will be subject to quite similar changes. To clear up the whole question, so far as this can be done by the gathering of observational evidence, it is only necessary to keep up a continual series of latitude determinations at several observatories. These determinations should show small variations similar in magnitude to the wabblings of the pole.

Let us now consider for a moment what is meant by the axis of the earth. It has long been known that the planet has in general the shape of a ball or sphere. That this is so can be seen at once from the way ships at sea disappear at the horizon. As they go farther and farther from us, we first lose sight of the hull, and then slowly and gradually the spars and sails seem to sink down into the ocean. This proves that the earth's surface is curved. That it is more or less like a sphere is evident from the fact that it always casts a round shadow in eclipses. Sometimes the earth passes between the sun and eclipsed moon. Then we see the earth's black shadow projected on the moon, which would otherwise be quite bright. This shadow has been observed in a very large number of such eclipses, and it has always been found to have a circular edge.

While, therefore, the earth is nearly a round ball, it must not be supposed that it is exactly spherical in form. We may disregard the small irregularities of its surface, for even the greatest mountains are insignificant in height when compared with the entire diameter of the earth itself. But even leaving these out of account, the earth is not perfectly spherical. We can describe it best as a flattened sphere. It is as though one were to press a round rubber ball between two smooth boards. It would be flattened at the top and bottom and bulged out in the middle. This is the shape of the earth. It is flattened at the poles and bulges out near the equator. The shortest straight line that can be drawn through the earth's centre and terminated by the flattened parts of its surface may be called the earth's axis of figure; and the two points where this axis meets the surface are called the poles of figure.

But the earth has another axis, called the axis of rotation. This is the one about which the planet turns once in a day, giving rise to the well-known phenomena called the rising and setting of sun, moon, and stars. For these motions of the heavenly bodies are really only apparent ones, caused by an actual motion of the observer on the earth. The observer turns with the earth on its axis, and is thus carried past the sun and stars.

This daily turning of the earth, then, takes place about the axis of rotation. Now, it so happens that all kinds of astronomical observations for the determination of latitude lead to values based on the rotation axis of the earth, and not on its axis of figure. We have seen how the earth's equator, from which we count our latitudes, is everywhere 90 degrees distant from the pole. But this pole is the pole of rotation, or the point at which the rotation axis pierces the earth's surface. It is not the pole of figure.

It is clear that the latitude of any observatory will remain constant only if the pole of figure and the rotation pole maintain absolutely the same positions relatively one to the other. These two poles are actually very near together; indeed, it was supposed for a very long time that they were absolutely coincident, so that there could not be any variations of latitude. But it now appears that they are separated slightly.

Strange to say, one of them is revolving about the other in a little curve. The pole of figure is travelling around the pole of rotation. The distance between them varies a little, never becoming greater than about fifty feet, and it takes about fourteen months to complete a revolution. There are some slight irregularities in the motion, but, in the main, it takes place in the manner here stated. In consequence of this rotation of the one pole about the other, the pole of figure is now on one side of the rotation pole and now on the opposite side, but it never travels continuously in one direction. Thus, as we have already seen, the sort of continuous motion required to explain the observed geological phenomena has not yet been found by astronomers.

Observations for the study of latitude variations have been made very extensively within recent years both in Europe and the United States. It has been found practically most advantageous to carry out simultaneous series of observations at two observatories situated in widely different parts of the earth, but having very nearly the same latitude. It is then possible to employ the same stars for observation in both places, whereas it would be necessary to use different sets of stars if there were much difference in the latitudes.

There is a special advantage in using the same stars in both places. We can then determine the small difference in latitude between the two participating observatories in a manner which will make it quite free from any uncertainty in our knowledge of the positions on the sky of the stars observed; for, strange as it may seem, our star-catalogues do not contain absolutely accurate numbers. Like all other data depending on fallible human observation, they are affected with small errors. But if we can determine simply the difference in latitude of the two observatories, we can discover from its variation the path in which the pole is moving. If, for instance, the observatories are separated by one-quarter the circumference of the globe, the pole will be moving directly toward one of them, when it is not changing its distance from the other one at all.

This method was used for seven years with good effect at the observatories of Columbia University in New York, and the Royal Observatory at Naples, Italy. For obtaining its most complete advantages it is, of course, better to establish several observing stations on about the same parallel of latitude. This was done in 1899 by the International Geodetic Association. Two stations are in the United States, one in Japan, and one in Sicily. We can, therefore, hope confidently that our knowledge as to the puzzling problem of polar motion will soon receive very material advancement.

SATURN'S RINGS

The death of James E. Keeler, Director of the Lick Observatory, in California (p. 32), recalls to mind one of the most interesting and significant of later advances in astronomical science. Only seven years have elapsed since Keeler made the remarkable spectroscopic observations which gave for the first time an ocular demonstration of the true character of those mysterious luminous rings surrounding the brilliant planet Saturn. His results have not yet been made sufficiently accessible to the public at large, nor have they been generally valued at their true worth. We consider this work of Keeler's interesting, because the problem of the rings has been a classic one for many generations; and we have been particular, also, to call it significant, because it is pregnant with the possibilities of newer methods of spectroscopic research, applied in the older departments of observational astronomy.

The troubles of astronomers with the rings began with the invention of the telescope itself. They date back to 1610, when Galileo first turned his new instrument to the heavens (p. 49). It may be imagined easily that the bright planet Saturn was among the very first objects scrutinized by him. His "powerful" instrument magnified only about thirty times, and was, doubtless, much inferior to our pocket telescopes of to-day. But it showed, at all events, that something was wrong with Saturn. Galileo put it, "*Ultimam planet am tergeminam observavi*" ("I have observed the furthest planet to be triple").

It is easy to understand now how Galileo's eyes deceived him. For a round luminous ball like Saturn, surrounded by a thin flat ring seen nearly edgewise, really looks as if it had two little attached appendages. Strange, indeed, it is to-day to read a scientific book so old that the planet Saturn could be called the "furthest" planet. But it was the outermost known in Galileo's day, and for nearly two centuries afterward. Not until 1781 did William Herschel discover Uranus (p. 59); and Neptune was not disclosed by the marvellous mathematical perception of Le Verrier until 1846 (p. 61).

Galileo's further observations of Saturn bothered him more and more. The planet's behavior became much worse as time went on. "Has Saturn devoured his children, according to the old legend?" he inquired soon afterward; for the changed positions of earth and planet in the course of their motions around the sun in their respective orbits had become such that the ring was seen quite edgewise, and was, therefore, perfectly invisible to Galileo's "optic tube." The puzzle remained unsolved by Galileo; it was left for another great man to find the true answer. Huygens, in 1656, first announced that the ring *is* a ring.

The manner in which this announcement was made is characteristic of the time; to-day it seems almost ludicrous. Huygens published a little pamphlet in 1656 called "*De Saturni Luna Observatio Nova*" or, "A New Observation of Saturn's Moon." He gave the explanation of what had been observed by himself and preceding astronomers in the form of a puzzle, or "logogriph." Here is what he had to say of the phenomenon in question:

"aaaaaaa ccccc d eeeee g h iiiiiii llll mm nnnnnnnnn oooo pp q rr s ttttt uuuuu."

It was not until 1659, three years later, in a book entitled "*Systema Saturnium*," that Huygens rearranged the above letters in their proper order, giving the Latin sentence:

"*Annulo cingitur, tenui plano, nusquam cohaerente, ad eclipticam inclinato.*" Translated into English, this sentence informs us that the planet "is girdled with a thin, flat ring, nowhere touching Saturn, and inclined to the ecliptic"!

This was a perfectly correct and wonderfully sagacious explanation of those complex and exasperatingly puzzling phenomena that had been too difficult for no less a person than Galileo himself. It was an explanation that *explained*. The reason for its preliminary announcement in the above manner must have been the following: Huygens was probably not quite sure of his ground in 1656, while three years afterward he had become quite certain. By the publication of the logogriph of 1656 he secured for himself the credit of what he had done. If any other astronomer had published the true explanation after 1656, Huygens could have proved his claim to priority by rearranging the letters of his puzzle. On the other hand, if further researches showed him that he was wrong, he would never have made known the true meaning of his logogriph, and would thus have escaped the ignominy of making an erroneous explanation. Thus, the method of announcement was comparable in ingenuity with the Huygenian explanation itself.

We are compelled to pass over briefly the entertaining history of subsequent observations of the ring, in order to explain the new work of Keeler and others. Cassini, about 1675, been able to show that the ring was double; that there are really two independent rings, with a distinct dark space between them. It was a case of wheels within wheels. To our own eminent countryman, W. C. Bond, of Cambridge, Mass., we owe the further discovery (Harvard College

Observatory, November, 1850) of the third ring. This is also concentric with the other two, and interior to them, but difficult to observe, because of its much smaller luminosity.

It is almost transparent, and the brilliant light of the planet's central ball is capable of shining directly through it. For this reason the inner ring is called the "gauze" or "crape" ring. If we add to the above details the fact that our modern large telescopes show slight irregularities in the surface of the rings, especially when seen edgewise, we have a brief statement of all that the telescope has been able to reveal to us since Galileo's time.

But of far greater interest than the mere fact of their existence is the important cosmic question as to the constitution, structure, and, above all, durability of the ring system. Astronomers often use the term "stability" with regard to celestial systems like the ring system of Saturn. By this they mean permanent durability. A system is stable if its various parts can continue in their present relationship to one another, without violating any of the known laws of astronomy. Whenever we study any collection of celestial objects, and endeavor to explain their motions and peculiarities, we always seek some explanation not inconsistent with the continued existence of the phenomena in question. For this there is, perhaps, no sufficient philosophical basis. Probably much of the great celestial procession is but a passing show, to be but for a moment in the endless vista of cosmic time.

However this may be, we are bound to assume as a working theory that Saturn has always had these rings, and will always have them; and it is for us to find out how this is possible. The problem has been attacked mathematically by various astronomers, including Laplace; but no conclusive mathematical treatment was obtained until 1857, when James Clerk Maxwell proved in a masterly manner that the rings could be neither solid nor liquid. He showed, indeed, that they would not last if they were continuous bodies like the planets. A big solid wheel would inevitably be torn asunder by any slight disturbance, and then precipitated upon the planet's surface. Therefore, the rings must be composed of an immense number of small detached particles, revolving around Saturn in separate orbits, like so many tiny satellites.

This mathematical theory of the ring system being once established, astronomers were more eager than ever to obtain a visual confirmation of it. We had, indeed, a sort of analogy in the assemblage of so-called "minor planets" (p. 64), which are known to be revolving around our sun in orbits situated between Mars and Jupiter. Some hundreds of these are known to exist, and probably there are countless others too small for us to see. Such a swarm of tiny particles of luminous matter would certainly give the impression of a continuous solid body, if seen from a distance comparable to that separating us from Saturn. But arguments founded on analogy are of comparatively little value.

Astronomers need direct and conclusive telescopic evidence, and this was lacking until Keeler made his remarkable spectroscopic observation in 1895. The spectroscope is a peculiar instrument, different in principle from any other used in astronomy; we study distant objects with it by analyzing the light they send us, rather than by examining and measuring the details of their visible surfaces. The reader will recall that according to the modern undulatory theory, light consists simply of a series of waves. Now, the nature of waves is very far from being understood in the popular mind. Most people, for instance, think that the waves of ocean consist of great masses of water rolling along the surface.

This notion doubtless arises from the behavior of waves when they break upon the shore, forming what we call surf. When a wave meets with an immovable body like a sand beach, the wave is broken, and the water really does roll upon the beach. But this is an exceptional case. Farther away from the shore, where the waves are unimpeded, they consist simply of particles of water moving straight up and down. None of the water is carried by mere wave-action away from the point over which it was situated at first.

Tides or other causes may move the water, but not simple wave-motion alone. That this is so can be proved easily. If a chip of wood be thrown overboard from a ship at sea it will be seen to rise and fall a long time on the waves, but it will not move. Similarly, wind-waves are often quite conspicuous on a field of grain; but they are caused by the individual grain particles moving up and down. The grain certainly cannot travel over the ground, since each particle is fast to its own stalk.

But while the particles do not travel, the wave-disturbance does. At times it is transmitted to a considerable distance from the point where it was first set in motion. Thus, when a stone is dropped into still water, the disturbance (though not the water) travels in ever-widening circles, until at last it becomes too feeble for us to perceive. Light is just such a travelling wave-disturbance. Beginning, perhaps, in some distant star, it travels through space, and finally the wave impinges on our eyes like the ocean-wave breaking on a sand beach. Such a light-wave affects the eye in some mysterious way. We call it "seeing."

The spectroscope (p. 21) enables us to measure and count the waves reaching us each second from any source of light. No matter how far away the origin of stellar light may be, the spectroscope examines the character of that light, and tells us the number of waves set up every second. It is this characteristic of the instrument that has enabled us to make some of the most remarkable observations of modern times. If the distant star is approaching us in space, more light-waves per second will reach us than we should receive from the same star at rest. Thus if we find from the spectroscope that there are too many waves, we know that the star is coming nearer; and if there are too few, we can conclude with equal certainty that the star is receding.

Keeler was able to apply the spectroscope in this way to the planet Saturn and to the ring system. The observations required dexterity and observational manipulative skill in a superlative degree. These Keeler had; and this work of his will always rank as a classic observation. He found by examining the light-waves from opposite sides of the planet that the luminous ball rotated; for one side was approaching us and the other receding. This observation was, of course, in accord with the known fact of Saturn's rotation on his axis. With regard to the rings, Keeler showed in the same way the existence of an axial rotation, which appears not to have been satisfactorily proved before, strange as it may seem. But the crucial point established by his spectroscope was that the interior part of the rings rotates *faster* than the exterior.

The velocity of rotation diminishes gradually from the inside to the outside. This fact is absolutely inconsistent with the motion of a solid ring; but it fits in admirably with the theory of a ring comprised of a vast assemblage of small separate particles. Thus, for the first time, astronomy comes into possession of an observational determination of the nature of Saturn's rings, and Galileo's puzzle is forever solved.

THE HELIOMETER

Astronomical discoveries are always received by the public with keen interest. Every new fact read in the great open book of nature is written eagerly into the books of men. For there exists a strong curiosity to ascertain just how the greater world is built and governed; and it must be admitted that astronomers have been able to satisfy that curiosity with no small measure of success. But it is seldom that we hear of the means by which the latest and most refined astronomical observations are effected. Popular imagination pictures the astronomer, as he doubtless once was, an aged gentleman, usually having a long white beard, and spending entire nights staring at the sky through a telescope.

But the facts to-day are very different. The working astronomer is an active man in the prime of life, often a young man. He wastes no time in star-gazing. His observations consist of exact measurements made in a precise, systematic, and almost business-like manner. A night's "watch" at the telescope is seldom allowed to exceed about three hours, since it is found that more continued exertions fatigue the eye and lead to less accurate results. To this, of course, there have been many notable exceptions, for endurance of sight, like any form of physical strength, differs greatly in different individuals. Astronomical research does not include "picking out" the constellations, and learning the Arabic names of individual stars. These things are not without interest; but they belong to astronomy's ancient history, and are of little value except to afford amusement and instruction to successive generations of amateurs.

Among the instruments for carefully planned measurements of precision the heliometer probably takes first rank. It is at once the most exquisitely accurate in its results, and the most fatiguing to the observer, of all the varied apparatus employed by the astronomer. The principle upon which its construction depends is very peculiar, and applies to all telescopes, even ordinary ones for terrestrial purposes. If part of a telescope lens be covered up with the hand, it will still be possible to see through the instrument. The glass lens at the end of the tube farthest from the observer's eye helps to magnify distant objects and make them seem nearer by gathering to a single point, or focus, a greater amount of their light than could be brought together by the far smaller lens in the unaided eye.

The telescope might very properly be likened to an enlarged eye, which can see more than we can, simply because it is bigger. If a telescope lens has a surface one hundred times as large as that of the lens in our eye, it will gather and bring to a focus one hundred times as much light from a distant object. Now, if any part of this telescope be covered, the remaining part will, nevertheless, gather and focus light just as though the whole lens were in action; only, there will be less light collected at the focus within the tube. The small lens at the telescope's eye-end is simply a magnifier to help our eye examine the image of any distant object formed at the focus by the large lens at the farther end of the instrument. For of this simple character is the operation of any telescope: the large glass lens at one end collects a distant planet's light, and brings it to a focus near the other end of the tube, where it forms a tiny picture of the planet, which, in turn, is examined with the little magnifier at the eye-end.

Having arrived at the fundamental principle that part of a lens will act in a manner similar to a whole one, it is easy to explain the construction of a heliometer. An ordinary telescope lens is sawed in half by means of a thin round metal disk revolved rapidly by machinery, and fed continually with emery and water at its edge. The cutting effect of emery is sufficient to make such a disk enter glass much as an ordinary saw penetrates wood. The two "semi-lenses," as they are called, are then mounted separately in metal holders. These are attached to one end of the heliometer, called the "head," in such a way that the two semi-lenses can slide side by side upon metal guides. This head is then fastened to one end of a telescope tube mounted in the usual way. The "head" end of the instrument is turned toward the sky in observing, and at the eye-end is placed the usual little magnifier we have already described.

The observer at the eye-end has control of certain rods by means of which he can push the semi-lenses on their slides in the head at the other end of the tube. Now, if he moves the semi-lenses so as to bring them side by side exactly, the whole arrangement will act like an ordinary telescope. For the semi-lenses will then fit together just as if the original glass had never been cut. But if the half-lenses are separated a little on their slides, each will act by itself. Being slightly separated, each will cover a different part of the sky. The whole affair acts as if the observer at the eye-end were looking through two telescopes at once. For each semi-lens acts independently, just as if it were a complete glass of only half the size.

Now, suppose there were a couple of stars in the sky, one in the part covered by the first semi-lens, and one in the part covered by the second. The observer would, of course, see both stars at once upon looking into the little magnifier at the eye-end of the heliometer.

We must remember that these stars will appear in the field of view simply as two tiny points of light. The astronomer, as we have said, is provided with a simple system of long rods, by means of which he can manipulate the semi-lenses while the observation is being made. If he slides them very slowly one way or the other, the two star-points in the field of view will be seen to approach each other. In this way they can at last be brought so near together that they will form but a single dot of light. Observation with the heliometer consists in thus bringing two star-images together, until at last they are superimposed one upon the other, and we see one image only. Means are provided by which it is then possible to measure the amount of separation of the two half-lenses. Evidently the farther asunder on the sky are the two stars under observation, the greater will be the separation of the semi-lenses necessary to make a single image of their light. Thus, measurement of the lenses' separation becomes a means of determining the separation of the stars themselves upon the sky.

The two slides in the heliometer head are supplied with a pair of very delicate measures or "scales" usually made of silver. These can be examined from the eye-end of the instrument by looking through a long microscope provided for this special purpose. Thus an extremely precise value is obtained both of the separation of the sliders and of the distance on the sky between the stars under examination. Measures made in this way with the heliometer are counted the most precise of astronomical observations.

Having thus described briefly the kind of observations obtained with the heliometer, we shall now touch upon their further utilization. We shall take as an example but one of their many uses—that one which astronomers consider the most important—the measurement of stellar distances. (See also p. 94.)

Even the nearest fixed star is almost inconceivably remote from us. And astronomers are imprisoned on this little earth; we cannot bridge the profound distance separating us from the stars, so as to use direct measurement with tape-line or surveyor's chain. We are forced to have recourse to some indirect method. Suppose a certain star is suspected, on account of its brightness, or for some other reason, of being near us in space, and so giving a favorable opportunity for a determination of distance. A couple of very faint stars are selected close by. These, on account of their faintness, the astronomer may regard as quite immeasurably far away. He then determines with his heliometer the exact position on the sky of the bright star with respect to the pair of faint ones. Half a year is then allowed to pass. During that time the earth has been swinging along in its annual path or orbit around the sun. Half a year will have sufficed to carry the observer on the earth to the other side of that path, and he is then 185,000,000 miles away from his position at the first observation.

Another determination is made of the bright star's position as referred to the two faint ones. Now, if all these stars were equally distant, their relative positions at the second observation would be just the same as at the former one. But if, as is very probable, the bright star is very much nearer us than are the two faint ones, we shall obtain a different position from our second observation. For the change of 185,000,000 miles in the observer's location will, of course, affect the direction in which we see the near star, while it will leave the distant ones practically unchanged. Without entering into technical details, we may say that from a large number of observations of this kind, we can obtain the distance of the bright star by a process of calculation. The only essential is to have an instrument that can make the actual observations of position accurately enough; and in this respect the heliometer is still unexcelled.

OCCULTATIONS

Scarcely anyone can have watched the sky without noticing how different is the behavior of our moon from that of any other object we can see. Of course, it has this in common with the sun and stars and planets, that it rises in the eastern horizon, slowly climbs the dome of the sky, and again goes down and sets in the west. This motion of the heavenly bodies is known to be an apparent one merely, and caused by the turning of our own earth upon its axis. A man standing upon the earth's surface can look up and see above him one-half the great celestial vault, gemmed with twinkling stars, and bearing, perhaps, within its ample curve one or two serenely shining planets and the lustrous moon. But at any given moment the observer can see nothing of the other half of the heavenly sphere. It is beneath his feet, and concealed by the solid bulk of the earth.

The earth, however, is turning on an axis, carrying the observer with it. And so it is continually presenting him to a new part of the sky. At any moment he sees but a single half-sphere; yet the very next instant it is no longer the same; a small portion has passed out of sight on one side by going down behind the turning earth, while a corresponding new section has come into view on the opposite side. It is this coming into view that we call the rising of a star; and the corresponding disappearance on the other side is called setting. Thus rising and setting are, of course, due entirely to a turning of the earth, and not at all to actual motions of the stars; and for this reason, all objects in the sky, without exception, must rise and set again. But the moon really has a motion of its own in addition to this apparent one caused by the earth's rotation.

Somewhere in the dawn of time early watchers of the stars thought out those fancied constellations that survive even down to our own day. They placed the mighty lion, king of beasts, upon the face of night, and the great hunter, too, armed with club and dagger, to pursue him. Among these constellations the moon threads her destined way, night after night, so rapidly that the unaided eye can see that she is moving. It needs but little power of fancy's magic to recall from the dim past a picture of some aged astronomer graving upon his tablets the Records of the Moon. "To-night she is near the bright star in the eye of the Bull." And again: "To-night she rides full, and near to the heart of the Virgin."

And why does the moon ride thus through the stars of night? Modern science has succeeded in disentangling the intricacies of her motion, until to-day only one or two of the very minutest details of that motion remain unexplained. But it has been a hard problem. Someone has well said that lunar theory should be likened to a lofty cliff upon whose side the intellectual giants among men can mark off their mental stature, but whose height no one of them may ever hope to scale.

But for our present purpose it is unnecessary to pursue the subject of lunar motion into its abstruser details. To understand why the moon moves rapidly among the stars, it is sufficient to remember that she is whirling quickly round the earth, so as to complete her circuit in a little less than a month. We see her at all times projected upon the distant background of the sky on which are set the stellar points of light, as though intended for beacons to mark the course pursued by moon and planets. The stars themselves have no such motions as the moon; situated at a distance almost inconceivably great, they may, indeed, be travellers through empty space, yet their journeys shrink into insignificance as seen from distant earth. It requires the most delicate instruments of the astronomer to so magnify the tiny displacements of the stars on the celestial vault that they may be measured by human eyes.

Let us again recur to our supposed observer watching the moon night after night, so as to record the stars closely approached by her. Why should he not some time or other be surprised by an approach so close as to amount apparently to actual contact? The moon covers quite a large surface on the sky, and when we remember the nearly countless numbers of the stars, it would, indeed, be strange if there were not some behind the moon as well as all around her.

A moment's consideration shows that this must often be the case; and in fact, if the moon's advancing edge be scrutinized carefully through a telescope, small stars can be seen frequently to disappear behind it. This is a most interesting observation. At first we see the moon and star near each other in the telescope's field of view. But the distance between them lessens perceptibly, even quickly, until at last, with a startling suddenness, the star goes out of sight behind the moon. After a time, ranging from a few moments to, perhaps, more than an hour, the moon will pass, and we can see the star suddenly reappear from behind the other edge.

These interesting observations, while not at all uncommon, can be made only very rarely by naked-eye astronomers. The reason is simple. The moon's light is so brilliant that it fairly overcomes the stars whenever they are at all near, except in the case of very bright ones. Small stars that can be followed quite easily up to the moon's edge in a good telescope, disappear from a naked-eye view while the moon is still a long distance away. But the number of very bright stars is comparatively small, so that it is quite unusual to find anyone not a professional astronomer who has actually seen one of these so-called "occultations." Moreover, most people are not informed in advance of the occurrence of an opportunity to make such observations, although they can be predicted quite easily by the aid of astronomical calculations. Sometimes we have occultations of planets, and these are the most interesting of all. When the moon passes between us and one of the larger planets, it is worth while to observe the phenomenon even without a telescope.

Up to this point we have considered occultations chiefly as being of interest to the naked-eye astronomer. Yet occultations have a real scientific value. It is by their means that we can, perhaps, best measure the moon's size. By noting with a telescope the time of disappearance and reappearance of known stars, astronomers can bring the direct measurement of the moon's diameter within the range of their numerical calculations. Sometimes the moon passes over a condensed cluster of stars like the Pleiades. The results obtainable on these occasions are valuable in a very high degree, and contribute largely to making precise our knowledge of the lunar diameter.

There is another thing of scientific interest about occultations, though it has lost some of its importance in recent years. When such an event has been observed, the agreement of the predicted time with that actually recorded by the astronomer offers a most searching test of the correctness of our theory of lunar motion. We have already called attention to the great inherent difficulty of this theory. It is easy to see that by noting the exact instant of disappearance of a star at a place on the earth the latitude and longitude of which are known, we can both check our calculations and gather material for improving our theory. The same principle can be used also in the converse direction. Within the limits of precision imposed by the state of our knowledge of lunar theory, we can utilize occultations to help determine the position on the earth of places whose longitude is unknown. It is a very interesting bit of history that the first determination of the longitude of Washington was made by means of occultations, and that this early determination led to the founding of the United States Naval Observatory.

On March 28, 1810, Mr. Pitkin, of Connecticut, reported to the House of Representatives on "laying a foundation for the establishment of a first meridian for the United States, by which a further dependence on Great Britain or any other foreign nation for such meridian may be entirely removed." This report was the result of a memorial presented by one William Lambert, who had deduced the longitude of the Capitol from an occultation observed October 20, 1804. Various proceedings were had in Congress and in committee, until at last, in 1821, Lambert was appointed "to make astronomical observations by lunar occultations of fixed stars, solar eclipses, or any approved method adapted to ascertain the longitude of the Capitol from Greenwich." Lambert's reports were made in 1822 and 1823, but ten years passed before the establishment of a formal Naval Observatory under Goldsborough, Wilkes, and Gilliss. But to Lambert belongs the honor of having marked out the first fundamental official meridian of longitude in the United States.

MOUNTING GREAT TELESCOPES

There are many interesting practical things about an astronomical observatory with which the public seldom has an opportunity to become acquainted. Among these, perhaps, the details connected with setting up a great telescope take first rank. The writer happened to be present at the Cape of Good Hope Observatory when the photographic equatorial telescope was being mounted, and the operation of putting it in position may be taken as typical of similar processes elsewhere. (See also p. 86.)

Forty-Inch Telescope, Yerkes Observatory,
University of Chicago.

In the first place, it is necessary to explain what is meant by an "equatorial" telescope. One of the chief difficulties in making ordinary observations arises from the rising and setting of the stars. They are all apparently moving across the face of the sky, usually climbing up from the eastern horizon, only to go down again and set in the west. If, therefore, we wish to scrutinize any given object for a considerable time, we must move the telescope continuously so as to keep pace with the motion of the heavens. For this purpose, the tube must be attached to axles, so that it can be turned easily in any direction. The equatorial mounting is a device that permits the telescope to be thus aimed at any part of the sky, and at the same time facilitates greatly the operation of keeping it pointed correctly after a star has once been brought into the field of view.

To understand the equatorial mounting it is necessary to remember that the rising and setting motions of the heavenly bodies are apparent ones only, and due in reality to the turning of the earth on its own axis. As the earth goes around, it carries observer, telescope, and observatory past the stars fixed upon the distant sky. Consequently, to keep a telescope pointed continuously at a given star, it is merely necessary to rotate it steadily backward upon a suitable axis just fast enough to neutralize exactly the turning of our earth.

By a suitable axis for this purpose we mean one so mounted as to be exactly parallel to the earth's own axis of rotation. A little reflection shows how simply such an arrangement will work. All the heavenly bodies may be regarded, for practical purposes, as excessively remote in comparison with the dimensions of our earth. The entire planet shrinks into absolute insignificance when compared with the distances of the nearest objects brought under observation by astronomers. It follows that if we have our telescope attached to such a rotation-axis as we have described, it will be just the same for purposes of observation as though the telescope's axis were not only parallel to the earth's axis, but actually coincident with it. The two axes may be separated by a distance equal to that between the earth's surface and its centre; but, as we have said, this distance is insignificant so far as our present object is concerned.

There is another way to arrive at the same result. We know that the stars in rising and setting all seem to revolve about the pole star, which itself seems to remain immovable. Consequently, if we mount our telescope so that it can turn about an axis pointing at the pole, we shall be able to neutralize the rotation of the stars by simply turning the telescope about the axis at the proper speed and in the right direction. Astronomical considerations teach us that an axis thus pointing at the pole will be parallel to the earth's own axis. Thus we arrive at the same fundamental principle for mounting an astronomical telescope from whichever point of view we consider the subject.

Every large telescope is provided with such an axis of rotation; and for the reason stated it is called the "polar axis." The telescope itself is then called an "equatorial." The advantage of this method of mounting is very evident. Since we can follow the stars' motions by turning the telescope about one axis only, it becomes a very simple matter to accomplish this turning automatically by means of clock-work.

The "following" of a star being thus provided for by the device of a polar axis, it is, of course, also necessary to supply some other motion so as to enable us to aim the tube at any point in the heavens. For it is obvious that if it were rigidly attached to the polar axis, we could, indeed, follow any star that happened to be in the field of view, but we could not change this field of view at will so as to observe other stars or planets. To accomplish this, the telescope is attached to the polar axis by means of a pivot. By turning the telescope around its polar axis, and also on this pivot, we can find any object in the heavens; and once found, we can leave to the polar axis and its automatic clock-work the task of keeping that object before the observer's eye.

In setting up the Cape of Good Hope instrument the astronomers were obliged to do a large part of the work of adjustment personally. Far away from European instrument-makers, the parts of the mounting and telescope had to be "assembled," or put together, by the astronomers of the Cape Observatory. A heavy pier of brick and masonry had been prepared in advance. Upon this was placed a massive iron base, intended to support the superstructure of polar axis and telescope. This base rested on three points, one of which could be screwed in and out, so as to tilt the whole affair a little forward or backward. By means of this screw we effected the final adjustment of the polar axis to exact parallelism with that of the earth. Other screws were provided with which the base could be twisted a little horizontally either to the right or left. Once set up in a position almost correct, it was easy to perfect the adjustment by the aid of these screws.

Afterward the tube and lenses were put in place, and the clock properly attached inside the big cast-iron base. This clock-work looked more like a piece of heavy machinery than a delicate clock mechanism. But it had heavy work to do, carrying the massive telescope with its weighty lenses, and needed to be correspondingly strong. It had a driving-weight of about 2,000 pounds, and was so powerful that turning the telescope affected it no more than the hour-hand of an ordinary clock affects the mechanism within its case.

The final test of the whole adjustment consisted in noting whether stars once brought into the telescopic field of view could be maintained there automatically by means of the clock. This object having been attained successfully, the instrument stood ready to be used in the routine business of the observatory.

Before leaving the subject of telescope-mountings, we must mention the giant instrument set up at the Paris Exposition of 1900. The project of having a *Grande Lunette* had been hailed by newspapers throughout the world and by the general public in their customary excitable way. It was tremendously over-advertised; exaggerated notions of the instrument's powers were spread abroad and eagerly credited; the moon was to be dragged down, as it were, from its customary place in the sky, so near that we should be able almost to touch its surface. As to the planets—free license was given to the journalistic imagination, and there was no effective limitation to the magnificence of astronomical discovery practically within our grasp, beyond the necessity for printed space demanded by sundry wars, pestilences, and other mundane trifles.

Yerkes Observatory, University of Chicago.

Now, the present writer is very far from advocating a lessening of the attention devoted to astronomy. Rather would he magnify his office than diminish it. But let journalistic astronomy be as good an imitation of sober scientific truth as can be procured at space rates; let editors encourage the public to study those things in the science that are ennobling and cultivating to the mind; let there be an end to the frenzied effort to fabricate a highly colored account of alleged discoveries of yesterday, capable of masquerading to-day under heavy head-lines as News.

The manner in which the big telescope came to be built is not without interest, and shows that enterprise is far from dead, even in the old countries. A stock company was organized—we should call it a corporation—under the name *Société de l'Optique*. It would appear that shares were regularly put on the market, and that a prospectus, more or less alluring, was widely distributed. We may say at once that the investing public did not respond with obtrusive alacrity; but at all events, the promoters' efforts received sufficient encouragement to enable them to begin active work. From the very first a vigorous attempt was made to utilize both the resources of genuine science and the devices of quasi-charlatanry. It was announced that the public were to be admitted to look through the big glass (apparently at so much an eye), and many, doubtless, expected that the man in the street would be able to make personal acquaintance with the man in the moon. A telescopic image of the sun was to be projected on a big screen, and exhibited to a concourse of spectators assembled in rising tiers of seats within a great amphitheatre. And when clouds or other circumstances should prevent observing the planets or scrutinizing the sun, a powerful stereopticon was to be used. Artificial pictures of the wonders of heaven were to be projected on the screen, and the public would never be disappointed. It was arranged that skilled talkers should be present to explain all marvels: and, in short, financial profit was to be combined with machinery for advancing scientific discovery. Astronomers the world over were "circularized," asked to become shareholders, and, in default of that, to send lantern-slides or photographs of remarkable celestial objects for exhibition in the magic-lantern part of the show.

The project thus brought to the attention of scientific men three years ago did not have an attractive air. It savored too much of charlatanism. But it soon appeared that effective government sanction had been given to the enterprise; and, above all, that men of reputation were allowing the use of their names in connection with the affair. More important still, we learned that the actual construction had been undertaken by Gautier, of Paris, that finances were favorable, and that real work on parts of the instrument was to commence without delay.

Gautier is a first-class instrument-builder; he has established his reputation by constructing successfully several telescopes of very large size, including the *equatorial coudé* of the Paris Observatory, a unique instrument of especial complexity. The present writer believes that, if sufficient time and money were available, the *Grande Lunette* would stand a reasonable chance of success in the hands of such a man. And by a reasonable chance, we mean that there is a large enough probability of genuine scientific discovery to justify the necessary financial outlay. But the project should be divorced from its "popular" features, and every kind of advertising and charlatanism excluded with rigor.

As planned originally, and actually constructed, the *Grande Lunette* presents interesting peculiarities, distinguishing it from other telescopes. Previous instruments have been built on the principle of universal mobility. It is possible to move them in all directions, and thus bring any desired star under observation, irrespective of its position in the sky. But this general mobility offers great difficulties in the case of large and ponderous telescopes. Delicacy of adjustment is almost destroyed when the object to be adjusted weighs several tons. And the excessive weight of telescopes is not due to unavoidably heavy lenses alone. It is essential that the tube be long; and great length involves appreciable thickness of material, if stiffness and solidity are to remain unsacrificed. Length in the tube is necessitated by certain peculiar optical defects of all lenses, into the nature of which we shall not enter at present. The consequences of these defects can be rendered harmless only if the instrument is so arranged that the observer's eye is far from the other end of the tube. The length of a good telescope should be at least twelve times the diameter of its large lens. If the relative length can be still further increased, so much the better; for then the optical defects can be further reduced.

In the case of the Paris instrument a radical departure consists in making the tube of unprecedented length, 197 feet, with a lens diameter of 49¼ inches. This great length, while favorable optically, precludes the possibility of making the instrument movable in the usual sense. In fact, the entire tube is attached to a fixed horizontal base, and no attempt is made to change its position. Outside the big lens, and disconnected altogether from the telescope proper, is mounted a smooth mirror, so arranged that it can be turned in any direction, and thus various parts of the sky examined by reflection in the telescope.

While this method unquestionably has the advantage of leaving the optician quite free as to how long he will make his tube, it suffers from the compensating objection that a new optical surface is introduced into the combination, viz., the mirror. Any slight unavoidable imperfection in the polishing of its surface will infallibly be reproduced on a magnified scale in the image of a distant star brought before the observer's eye.

But it is not yet possible to pronounce definitely upon the merit of this form of instrument, since, as we have said, the maker has not been given time enough to try the idea to the complete satisfaction of scientific men. In the early part of August, 1900, when the informant of the present writer left Paris, after serving as a member of the international jury for judging instruments of precision at the Exposition, the condition of the *Grande Lunette* was as follows: Two sets of lenses had been contemplated, one intended for celestial photography, and the other to be used for ordinary visual observation. Only the photographic lenses had been completed, however, and for this reason the public could not be permitted to look through the instrument. The photographic lenses were in place in the tube, but at that time their condition was such that, though some photographs had been obtained, it was not thought advisable to submit them to the jury. Consequently, the *Lunette* did not receive a prize. Since that time various newspapers have reported wonderful results from the telescope; but, disregarding the fusillade from the sensational press, we may sum up the present state of affairs very briefly. Gautier is still experimenting; and, given sufficient time and money, he may succeed in producing what astronomers hope for—an instrument capable of advancing our knowledge, even if that advance be only a small one.

THE ASTRONOMER'S POLE

The pole of the frozen North is not the only pole sought with determined effort by more than one generation of scientific men. Up in the sky astronomers have another pole which they are following up just as vigorously as ever Arctic explorer struggled toward the difficult goal of his terrestrial journeying. The celestial pole is, indeed, a fundamentally important thing in astronomical science, and the determination of its exact position upon the sky has always engaged the closest attention of astronomers. Quite recently new methods of research have been brought to bear, promising a degree of success not hitherto attained in the astronomers' pursuit of their pole.

In the first place, we must explain what is meant by the celestial pole. We have already mentioned the poles of the earth (p. 136). Our planet turns once daily upon an axis passing through its centre, and it is this rotation that causes all the so-called diurnal phenomena of the heavens. Rising and setting of sun, moon, and stars are simply results of this turning of the earth. Heavenly bodies do not really rise; it is merely the man on the earth who is turned round on an axis until he is brought into a position from which he can see them. The terrestrial poles are those two points on the earth's surface where it is pierced by the rotation axis of the planet. Now we can, if we choose, imagine this axis lengthened out indefinitely, further and further, until at last it reaches the great round vault of the sky. Here it will again pierce out two polar points; and these are the celestial poles.

The whole thing is thus quite easy to understand. On the sky the poles are marked by the prolongation of the earth's axis, just as on the earth the poles are marked by the axis itself. And this explains at once why the stars seem nightly to revolve about the pole. If the observer is being turned round the earth's axis, of course it will appear to him as if the stars were rotating around the same axis in the opposite direction, just as houses and fields seem to fly past a person sitting in a railway train, unless he stops to remember that it is really himself who is in motion, and not the trees and houses.

The existence of such a centre of daily motions among the stars once recognized, it becomes of interest to ascertain whether the centre itself always retains precisely the same position in the sky. It was discovered as early as the time of Hipparchus (p. 39) that such is not the case, and that the celestial pole is subject to a slow motion among the stars on the sky. If a given star were to-day situated exactly at the pole, it would no longer be there after the lapse of a year's time; for the pole would have moved away from it.

This motion of the pole is called precession. It means that certain forces are continually at work, compelling the earth's axis to change its position, so that the prolongation of that axis must pierce the sky at a point which moves as time goes on. These forces are produced by the gravitational attractions of the sun, moon, and planets upon the matter composing our earth. If the earth were perfectly spherical in shape, the attractions of the other heavenly bodies would not affect the direction of the earth's rotation-axis in the least. But the earth is not quite globular in form; it is flattened a little at the poles and bulges out somewhat at the equator. (See p. 135.)

This protuberant matter near the equator gives the other bodies in the solar system an opportunity to disturb the earth's rotation. The general effect of all these attractions is to make the celestial pole move upon the sky in a circle having a radius of about 23½ degrees; and it requires 25,800 years to complete a circuit of this precessional cycle. One of the most striking consequences of this motion will be the change of the polar star. Just at present the bright star Polaris in the constellation of the Little Bear is very close to the pole. But after the lapse of sufficient ages the first-magnitude star Vega of the constellation Lyra will in its turn become Guardian of the Pole.

It must not be supposed, however, that the motion of the pole proceeds quite uniformly, and in an exact circle; the varying positions of the heavenly bodies whose attractions cause the phenomena in question are such as to produce appreciable divergencies from exact circular motion. Sometimes the pole deviates a little to one side of the precessional circle, and sometimes it deviates on the other side. The final result is a sort of wavy line, half on one side and half on the other of an average circular curve. It takes only nineteen years to complete one of these little waves of polar motion, so that in the whole precessional cycle of 25,800 years there are about 1,400 indentations. This disturbance of the polar motion is called by astronomers nutation.

The first step in a study of polar motion is to devise a method of finding just where the pole is on any given date. If the astronomer can ascertain by observational processes just where the pole is among the stars at any moment, and can repeat his observations year after year and generation after generation, he will possess in time a complete chart of a small portion at least of the celestial pole's vast orbit. From this he can obtain necessary data for a study of the mathematical theory of attractions, and thus, perhaps, arrive at an explanation of the fundamental laws governing the universe in which we live.

The instrument which has been used most extensively for the study of these problems is the transit (p. 118) or the "meridian circle." This latter consists of a telescope firmly attached to a metallic axis about which it can turn. The axis itself rests on massive stone supports, and is so placed that it points as nearly as possible in an east-and-west direction. Consequently, when the telescope is turned about its axis, it will trace out on the sky a great circle (the meridian) which passes through the north and south points of the horizon and the point directly overhead. The instrument has also a metallic circle very firmly fastened to the telescope and its axis. Let into the surface of this circle is a silver disk upon which are engraved a series of lines or graduations by means of which it is possible to measure angles.

Observers with the meridian circle begin by noting the exact instant when any given star passes the centre of the field of view of the telescope. This centre is marked with a cross made by fastening into the focus some pieces of ordinary spider's web, which give a well-marked, delicate set of lines, even under the magnifying power of the telescope's eye-piece. In addition to thus noting the time when the star crosses the field of the telescope, the astronomer can measure by means of the circle, how high up it was in the sky at the instant when it was thus observed.

If the telescope of the meridian circle be turned toward the north, and we observe stars close to the pole, it is possible to make two different observations of the same star. For the close polar stars revolve in such small circles around the pole of the heavens that we can observe them when they are on the meridian either above the pole or below it. Double observations of this class enable us to obtain the elevation of the pole above the horizon, and to fix its position with respect to the stars.

Now, there is one very serious objection to this method. In order to secure the two necessary observations of the same star, it is essential to be stationed at the instrument at two moments of time separated by exactly twelve hours; and if one of the observations occurs in the night, the other corresponding observation will occur in daylight.

It is a fact not generally known that the brighter stars can be seen with a telescope, even when the sun is quite high above the horizon. Unfortunately, however, there is only one star close to the pole which is bright enough to be thus observed in daylight—the polar star already mentioned under the name Polaris. The fact that we are thus limited to observations of a single star has made it difficult even for generations of astronomers to accumulate with the meridian circle a very large quantity of observational material suitable for the solution of our problem.

The new method of observation to which we have referred above consists in an application of photography to the polar problem. If we aim at the pole a powerful photographic telescope, and expose a photographic plate throughout the entire night, we shall find that all stars coming within the range of the plate will mark out little circles or "trails" upon the developed negative. It is evident that as the stars revolve about the pole on the sky, tracing out their daily circular orbits, these same little circles must be reproduced faithfully upon the photographic plate. The only condition is that the stars shall be bright enough to make their light affect the sensitive gelatine surface.

But even if observations of this kind are continued throughout all the hours of darkness, we do not obtain complete circles, but only those portions of circles traced out on the sky between sunset and sunrise. If the night is twelve hours in length, we get half-circles on the plate; if it is eighteen hours long, we get circles that lack only one-quarter of being complete. In other words, we get a series of circular arcs, one corresponding to each close polar star. There are no fewer than sixteen stars near enough to the pole to come within the range of a photographic plate, and bright enough to cause measurable impressions upon the sensitive surface. The fact that the circular arcs are not complete circles does not in the least prevent our using them for ascertaining the position of their common centre; and that centre is the pole. Moreover, as the arcs are distributed at all sorts of distances from the pole and in all directions, corresponding to the accidental positions of the stars on the sky, we have a state of affairs extremely favorable to the accurate determination of the pole's place among the stars by means of microscopic measurements of the plate.

It will be perceived that this method is extremely simple, and, therefore, likely to be successful; though its simplicity is slightly impaired by the phenomenon known to astronomers as "atmospheric refraction." The rays of light coming down to our telescopes from a distant star must pass through the earth's atmosphere before they reach us; and in passing thus from the nothingness of outer space into the denser material of the air, they are bent out of their straight course. The phenomenon is analogous to what we see when we push a stick down through the surface of still water; we notice that the stick appears to be bent at the point where it pierces the surface of the water; and in just the same way the rays of light are bent when they pierce into the air. Fortunately, the mathematical theory of this atmospheric bending of light is well understood, so that it is possible to remove the effects of refraction from our results by a process of calculation. In other words, we can transform our photographic measures into what they would have been if no such thing as atmospheric refraction existed. This having been done, all the arcs on the plate should be exactly circular, and their common centre should be the position of the pole among the stars on the night when the photograph was made.

It is possible to facilitate the removal of refraction effects very much by placing our photographic telescope at some point on the earth situated in a very high latitude. The elevation of the pole above the horizon is greatest in high latitudes. Indeed, if Arctic voyagers could ever reach the pole of the earth they would see the pole of the heavens directly overhead. Now, the higher up the pole is in the sky, the less will be the effects of atmospheric refraction; for the rays of light will then strike the atmosphere in a direction nearly perpendicular to its surface, which is favorable to diminishing the amount of bending.

There is also another very important advantage in placing the telescope in a high latitude; in the middle of winter the nights are very long there; if we could get within the Arctic. Circle itself, there would be nights when the hours of darkness would number twenty-four, and we could substitute complete circles for our broken arcs. This would, indeed, be most favorable from the astronomical point of view; but the essential condition of convenience for the observer renders an expedition to the frozen Arctic regions unadvisable.

But it is at least possible to place the telescope as far north as is consistent with retaining it within the sphere of civilized influences. We can put it in that one of existing observatories on the earth which has the highest latitude; and this is the observatory of Helsingfors, in Finland, which belongs to a great university, is manned by competent astronomers, and has a latitude greater than 60 degrees.

Dr. Anders Donner, Director of the Helsingfors Observatory, has at its disposal a fine photographic telescope, and with this some preliminary experimental "trail" photographs were made in 1895. These photographs were sent to Columbia University, New York, and were there measured under the writer's direction. Calculations based on these measures indicate that the method is promising in a very high degree; and it was, therefore, decided to construct a special photographic telescope better adapted to the particular needs of the problem in hand.

The desirability of a new telescope arises from the fact that we wish the instrument to remain absolutely unmoved during all the successive hours of the photographic exposure. It is clear that if the telescope moves while the stars are tracing out their little trails on the plate, the circularity of the curves will be disturbed. Now, ordinary astronomical telescopes are always mounted upon very stable foundations, well adapted to making the telescope stand still; but the polar telescope which we wish to use in a research fundamental to the entire science of astronomy ought to possess immobility and stability of an order higher than that required for ordinary astronomical purposes.

It is a remarkable peculiarity of the instrument needed for the new trail photographs that it is never moved at all. Once pointed at the pole, it is ready for all the observations of successive generations of astronomers. It should have no machinery, no pivots, axes, circles, clocks, or other paraphernalia of the usual equatorial telescope. All we want is a very heavy stone pier, with a telescope tube firmly fastened to it throughout its entire length. The top of the pier having been cut to the proper angle of the pole's elevation, and the telescope cemented down, everything is complete from the instrumental side; and just such an instrument as this is now ready for use at Helsingfors.

The late Miss Catharine Wolfe Bruce, of New York, was much interested in the writer's proposed polar investigations, and in October, 1898, she contributed funds for the construction of the new telescope, and the Russian authorities have generously undertaken the expense of a building to hold the instrument and the granite foundation upon which it rests. Photographs are now being secured with the new instrument, and they will be sent to Columbia University, New York, for measurement and discussion. It is hoped that they will carry out the promise of the preliminary photographs made in 1895 with a less suitable telescope of the ordinary form.

THE MOON HOAX

The public attitude toward matters scientific is one of the mysteries of our time. It can be described best by the single word, Credulity; simple, absolute credulity. Perfect confidence is the most remarkable characteristic of this unbelieving age. No charlatan, necromancer, or astrologer of three centuries ago commanded more respectful attention than does his successor of to-day.

Any person can be a scientific authority; he has but to call himself by that title, and everyone will give him respectful attention. Numerous instances can be adduced from the experience of very recent years to show how true are these remarks. We have had the Keeley motor and the liquid-air power schemes for making something out of nothing. Extracting gold from sea-water has been duly heralded on scientific authority as an easy source of fabulous wealth for the million. Hard-headed business men not only believe in such things, but actually invest in them their most valued possession, capital. Venders of nostrums and proprietary medicines acquire wealth as if by magic, though it needs but a moment's reflection to realize that these persons cannot possibly be in possession of any drugs, or secret methods of compounding drugs, that are unknown to scientific chemists.

If the world, then, will persistently intrust its health and wealth into the safe-keeping of charlatans, what can we expect when things supposedly of far less value are at stake? The famous Moon Hoax, as we now call it, is truly a classic piece of lying. Though it dates from as long ago as 1835, it has never had an equal as a piece of "modern" journalism. Nothing could be more useful than to recall it to public attention at least once every decade; for it teaches an important lesson that needs to be iterated again and again.

On November 13, 1833, Sir John Herschel embarked on the Mountstuart Elphinstone, bound for the Cape of Good Hope. He took with him a collection of astronomical instruments, with which he intended to study the heavens of the southern hemisphere, and thus extend his father's great work to the south polar stars. An earnest student of astronomy, he asked no better than to be left in peace to seek the truth in his own fashion. Little did he think that his expedition would be made the basis for a fabrication of alleged astronomical discoveries destined to startle a hemisphere. Yet that is precisely what happened. Some time about the middle of the year 1835 the New York *Sun* began the publication of certain articles, purporting to give an account of "Great Astronomical Discoveries, lately made by Sir John Herschel at the Cape of Good Hope." It was alleged that these articles were taken from a supplement to the Edinburgh *Journal of Science*; yet there is no doubt that they were manufactured entirely in the United States, and probably in New York.

The hoax begins at once in a grandiloquent style, calculated to attract popular attention, and well fitted to the marvels about to be related. Here is an introductory remark, as a specimen: "It has been poetically said that the stars of heaven are the hereditary regalia of man as the intellectual sovereign of the animal creation. He may now fold the zodiac around him with a loftier consciousness of his mental supremacy." Then follows a circumstantial and highly plausible account of the manner in which early and exclusive information was obtained from the Cape. This was, of course, important in order to make people believe in the genuineness of the whole; but we pass at once to the more interesting account of Herschel's supposed instrument.

Nothing could be more skilful than the way in which an air of truth is cast over the coming account of marvellous discoveries by explaining in detail the construction of the imaginary Herschelian instrument. Sir John is supposed to have had an interesting conversation in England "with Sir David Brewster, upon the merits of some ingenious suggestion by the latter, in his article on optics in the Edinburgh Encyclopædia (p. 644), for improvements in the Newtonian reflectors." The exact reference to a particular page is here quite delightful. After some further talk, "the conversation became directed to that all-invincible enemy, the paucity of light in powerful magnifiers. After a few moments' silent thought, Sir John diffidently inquired whether it would not be possible to effect a *transfusion of artificial light through the focal object of vision*! Sir David, somewhat startled at the originality of the idea, paused awhile, and then hesitatingly referred to the refrangibility of rays, and the angle of incidence.... Sir John continued, 'Why cannot the illuminated microscope, say the hydro-oxygen, be applied to render distinct, and, if necessary, even to magnify the focal object?' Sir David sprang from his chair in an ecstasy of conviction, and leaping half-way to the ceiling, exclaimed, 'Thou art the man.' "This absurd imaginary conversation contains nothing but an assemblage of optical jargon, put together without the slightest intention of conveying any intelligible meaning to scientific people. Yet it was well adapted to deceive the public; and we should not be surprised if it would be credited by many newspaper readers to-day.

The authors go on to explain how money was raised to build the new instrument, and then describe Herschers embarkation and the difficulties connected with transporting his gigantic machines to the place selected for the observing station. "Sir John accomplished the ascent to the plains by means of two relief teams of oxen, of eighteen each, in about four days, and, aided by several companies of Dutch boors [*sic*], proceeded at once to the erecting of his gigantic fabric." The place really selected by Herschel cannot be described better than in his own words, contained in a genuine letter dated January 21, 1835: "A perfect paradise in rich and magnificent mountain scenery, sheltered from all winds.... I must reserve for my next all description of the gorgeous display of flowers which adorn this splendid country, as well as the astonishing brilliancy of the constellations." The author of the hoax could have had no knowledge of Herschers real location, as described in this letter.

The present writer can bear witness to the correctness of Herschel's words. Feldhausen is truly an ideal secluded spot for astronomical study. A small obelisk under the sheer cliff of far-famed Table Mountain now marks the site of the great reflecting telescope. Here Herschel carried on his scrutiny of the Southern skies. He observed 1,202 double stars and 1,708 nebulæ and clusters, of which only 439 were already known. He studied the famous Magellanic clouds, and made the first careful drawings of the "keyhole" nebula in the constellation Argo.

Very recent researches of the present royal astronomer at the Cape have shown that changes of import have certainly taken place in this nebula since Herschel's time, when a sudden blazing up of the wonderful star Eta Argus was seen within the nebula. This object has, perhaps, undergone more remarkable changes of light than any other star in the heavens. It is as though there were some vast conflagration at work, now blazing into incandescence, and again sinking almost into invisibility. In 1843 Maclear estimated the brilliancy of Eta to be about equal to that of Sirius, the brightest star in the whole sky. Later it diminished in light, and cannot be seen to-day with the naked eye, though the latest telescopic observations indicate that it is again beginning to brighten.

Such was Herschel's quiet study of his beloved science, in glaring contrast to the supposed discoveries of the "Hoax." Here are a few things alleged to have been seen on the moon. The first time the instrument was turned upon our satellite "the field of view was covered throughout its entire area with a beautifully distinct and even vivid representation of basaltic rock." There were forests, too, and water, "fairer shores never angels coasted on a tour of pleasure. A beach of brilliant white sand, girt with wild castellated rocks, apparently of green marble."

There was animal life as well; "we beheld continuous herds of brown quadrupeds, having all the external characteristics of the bison, but more diminutive than any species of the bos genus in our natural history." There was a kind of beaver, that "carries its young in its arms like a human being," and lives in huts. "From the appearance of smoke in nearly all of them, there is no doubt of its (the beaver's) being acquainted with the use of fire." Finally, as was, of course, unavoidable, human creatures were discovered. "Whilst gazing in a perspective of about half a mile, we were thrilled with astonishment to perceive four successive flocks of large-winged creatures, wholly unlike any kind of birds, descend with a slow, even motion from the cliffs on the western side, and alight upon the plain.... Certainly they were like human beings, and their attitude in walking was both erect and dignified."

We have not space to give more extended extracts from the hoax, but we think the above specimens will show how deceptive the whole thing was. The rare reprint from which we have extracted our quotations contains also some interesting "Opinions of the American Press Respecting the Foregoing Discovery."

The *Daily Advertiser* said: "No article, we believe, has appeared for years, that will command so general a perusal and publication. Sir John has added a stock of knowledge to the present age that will immortalize his name and place it high on the page of science." The *Mercantile Advertiser* said: "Discoveries in the Moon.—We commence to-day the publication of an interesting article which is stated to have been copied from the Edinburgh *Journal of Science*, and which made its first appearance here in a contemporary journal of this city. It appears to carry intrinsic evidence of being an authentic document." Many other similar extracts are given. The New York *Evening Post* did not fall into the trap. The *Evening Post's* remarks were as follows: "It is quite proper that the *Sun* should be the means of shedding so much light on the *Moon*. That there should be winged people in the moon does not strike us as more wonderful than the existence of such a race of beings on the earth; and that there does or did exist such a race rests on the evidence of that most veracious of voyagers and circumstantial of chroniclers, Peter Wilkins, whose celebrated work not only gives an account of the general appearance and habits of a most interesting tribe of flying Indians, but also of all those more delicate and engaging traits which the author was enabled to discover by reason of the conjugal relations he entered into with one of the females of the winged tribe."

We shall limit our extracts from the contemporary press to the few quotations here given, hoping that enough has been said to direct attention once more to that important subject, the Possibility of Being Deceived.

THE SUN'S DESTINATION

Three generations of men have come and gone since the Marquis de Laplace stood before the Academy of France and gave his demonstration of the permanent stability of our solar system. There was one significant fault in Newton's superbly simple conception of an eternal law governing the world in which we live. The labors of mathematicians following him had shown that the planets must trace out paths in space whose form could be determined in advance with unerring certainty by the aid of Newton's law of gravitation. But they proved just as conclusively that these planetary orbits, as they are called, could not maintain indefinitely the same shapes or positions. Slow indeed might be the changes they were destined to undergo; slow, but sure, with that sureness belonging to celestial science alone. And so men asked: Has this magnificent solar system been built upon a scale so grand, been put in operation subject to a law sublime in its very simplicity, only to change and change until at length it shall lose every semblance of its former self, and end, perhaps, in chaos or extinction?

Laplace was able to answer confidently, "No." Nor was his answer couched in the enthusiastic language of unbalanced theorists who work by the aid of imagination alone. Based upon the irrefragable logic of correct mathematical reasoning, and clad in the sober garb of mathematical formulæ, his results carried conviction to men of science the world over. So was it demonstrated that changes in our solar system are surely at work, and shall continue for nearly countless ages; yet just as surely will they be reversed at last, and the system will tend to return again to its original form and condition. The objection that the Newtonian law meant ultimate dissolution of the world was thus destroyed by Laplace. From that day forward the law of gravitation has been accepted as holding sway over all phenomena visible within our planetary world.

The intricacies of our own solar system being thus illumined, the restless activity of the human intellect was stimulated to search beyond for new problems and new mysteries. Even more fascinating than the movements of our sun and planets are all those questions that relate to the clustered stellar congeries hanging suspended within the deep vault of night. Does the same law of gravitation cast its magic spell over that hazy cloud of Pleiades, binding them, like ourselves, with bonds indissoluble? Who shall answer, yes or no? We can only say that astronomers have as yet but stepped upon the threshold of the universe, and fixed the telescope's great eye upon that which is within.

Let us then begin by reminding the reader what is meant by the Newtonian law of gravitation. It appears all things possess the remarkable property of attracting or pulling each other. Newton declared that all substances, solid, liquid, or even gaseous—from the massive cliff of rock down to the invisible air—all matter can no more help pulling than it can help existing. His law further formulates certain conditions governing the manner in which this gravitational attraction is exerted; but these are mere matters of detail; interest centres about the mysterious fact of attraction itself. How can one thing pull another with no connecting link through which the pull can act? Just here we touch the point that has never yet been explained. Nature withholds from science her ultimate secrets. They that have pondered longest, that have descended farthest of all men into the clear well of knowledge, have done so but to sound the depths beyond, never touching bottom.

This inability of ours, to give a good physical explanation of gravitation, has led certain makers of paradoxes to doubt or even deny that there is any such thing. But, fortunately, we have a simple laboratory experiment that helps us. Unexplained it may ever remain, but that there can be attraction between physical objects connected by no visible link is proved by the behavior of an ordinary magnet. Place a small piece of steel or iron near a magnetized bar, and it will at once be so strongly attracted that it will actually fly to the magnet. Anyone who has seen this simple experiment can never again deny the possibility, at least, of the law of attraction as stated by Newton. Its possibility once admitted, the fact that it can predict the motions of all the planets, even down to their minutest details, transforms the possibility of its truth into a certainty as strong as any human certainty can ever be.

But this demonstration of Newton's law is limited strictly to the solar system itself. We may, indeed, reason by analogy, and take for granted that a law which holds within our immediate neighborhood is extremely likely to be true also of the entire visible universe. But men of science are loath to reason thus; and hence the fascination of researches in cosmic astronomy. Analogy points out the path. The astronomer is not slow to follow; but he seeks ever to establish upon incontrovertible evidence those truths which at first only his daring imagination had led him to half suspect.

If we are to extend the law of gravitation to the utmost, we must be careful to consider the law itself in its most complete form. A heavenly body like the sun is often said to govern the motions of its family of planets; but such a statement is not strictly accurate. The governing body is no despot; 'tis an abject slave of law and order, as much as the tiniest of attendant planets. The action of gravitation is mutual, and no cosmic body can attract another without being itself in turn subject to that other's gravitational action.

If there were in our solar system but two bodies, sun and planet, we should find each one pursuing a path in space under the influence of the other's attraction. These two paths or orbits would be oval, and if the sun and planet were equally massive, the orbits would be exactly alike, both in shape and size. But if the sun were far larger than the planet, the orbits would still be similar in form, but the one traversed by the larger body would be small. For it is not reasonable to expect a little planet to keep the big sun moving with a velocity as great as that derived by itself from the attraction of the larger orb.

Whenever the preponderance of the larger body is extremely great, its orbit will be correspondingly insignificant in size. This is in fact the case with our own sun. So massive is it in comparison with the planets that the orbit is too small to reveal its actual existence without the aid of our most refined instruments. The path traced out by the sun's centre would not fill a space as large as the sun's own bulk. Nevertheless, true orbital motion is there.

So we may conclude that as a necessary consequence of the law of gravitation every object within the solar system is in motion. To say that planets revolve about the sun is to neglect as unimportant the small orbit of the sun itself. This may be sufficiently accurate for ordinary purposes; but it is unquestionably necessary to neglect no factor, however small, if we propose to extend our reasoning to a consideration of the stellar universe. For we shall then have to deal with systems in which the planets are of a size comparable with the sun; and in such systems all the orbits will also be of comparatively equal importance.

Mathematical analysis has derived another fact from discussion of the law of gravitation which, perhaps, transcends in simple grandeur everything we have as yet mentioned. It matters not how great may be the number of massive orbs threading their countless interlacing curved paths in space, there yet must be in every cosmic system one single point immovable. This point is called the Centre of Gravity. If it should so happen that in the beginning of things, some particle of matter were situated at this centre, then would that atom ever remain unmoved and imperturbable throughout all the successive vicissitudes of cosmic evolution. It is doubtful whether the mind of man can form a conception of anything grander than such an immovable atom within the mysterious intricacies of cosmic motion.

But in general, we cannot suppose that the centres of gravity in the various stellar systems are really occupied by actual physical bodies. The centre may be a mere mathematical point in space, situated among the several bodies composing the system, but, nevertheless, endowed, in a certain sense, with the same remarkable property of relative immobility.

Having thus defined the centre of gravity in its relation to the constituent parts of any cosmic system, we can pass easily to its characteristic properties in connection with the inter-relation of stellar systems with one another. It can be proved mathematically that our solar system will pull upon distant stars just as though the sun and all the planets were concentrated into one vast sphere having its centre in the centre of gravity of the whole. It is this property of the centre of gravity which makes it pre-eminently important in cosmic researches. For, while we know that centre to be at rest relatively to all the planets in the system, it may, nevertheless, in its quality as a sort of concentrated essence of them all, be moving swiftly through space under the pull of distant stars. In that case, the attendant bodies will go with it—but they will pursue their evolutions within the system, all unconscious that the centre of gravity is carrying them on a far wider circuit.

What is the nature of that circuit? This question has been for many years the subject of earnest study by the clearest minds among astronomers. The greatest difficulty in the way is the comparatively brief period during which men have been able to make astronomical observations of precision. Space and time are two conceptions that transcend the powers of definition possessed by any man. But we can at least form a notion of how vast is the extent of time, if we remember that the period covered by man's written records is registered but as a single moment upon the great revolving dial of heaven's dome. One hundred and fifty years have elapsed since James Bradley built the foundations of modern sidereal astronomy upon his masterly series of observations at the Royal Observatory of Greenwich, in England. Yet so slowly do the movements of the stars unroll themselves upon the firmament, that even to this day no one of them has been seen by men to trace out more than an infinitesimal fraction of its destined path through the voids of space.

Travellers upon a railroad cannot tell at any given moment whether they are moving in a straight line, or whether the train is turning upon some curve of huge size. The St. Gothard railway has several so-called "corkscrew" tunnels, within which the rails make a complete turn in a spiral, the train finally emerging from the tunnel at a point almost vertically over the entrance. In this way the train is lifted to a higher level. Passengers are wont to amuse themselves while in these tunnels by watching the needle of an ordinary pocket-compass. This needle, of course, always points to the north; and as the train turns upon its curve, the needle will make a complete revolution. But the passenger could not know without the compass that the train was not moving in a perfectly straight line. Just so we passengers on the earth are unaware of the kind of path we are traversing, until, like the compass, the astronomer's instruments shall reveal to us the truth.

But as we have seen, astronomical observations of precision have not as yet extended through a period of time corresponding to the few minutes during which the St. Gothard traveller watches the compass. We are still in the dark, and do not know as yet whether mankind shall last long enough upon the earth to see the compass needle make its revolution. We are compelled to believe that the motion in space of our sun is progressing upon a curved path; but so far as precise observations allow us to speak, we can but say that we have as yet moved through an infinitesimal element only of that mighty curve. However, we know the point upon the sky toward which this tiny element of our path is directed, and we have an approximate knowledge of the speed at which we move.

More than a century ago Sir William Herschel was able to fix roughly what we call the apex of the sun's way in space, or the point among the stars toward which that way is for the moment directed. We say for the moment, but we mean that moment of which Bradley saw the beginning in 1750, and upon whose end no man of those now living shall ever look. Herschel found that a comparison of old stellar observations seemed to indicate that the stars in a certain part of the sky were opening out, as it were, and that the constellations in the opposite part of the heavens seemed to be drawing in, or becoming smaller. There can be but one reasonable explanation of this. We must be moving toward that part of the sky where the stars are separating. Just so a man watching a regiment of soldiers approaching, will see at first only a confused body of men; but as they come nearer, the individual soldiers will seem to separate, until at length each one is seen distinct from all the others.

Herschel fixed the position of the apex at a point in the constellation Hercules. The most recent investigations of Newcomb and others have, on the whole, verified Herschel's conclusions. With the intuitive power of rare genius, Herschel had been able to sift truth out of error. The observational data at his disposal would now be called rude, but they disclosed to the scrutiny of his acute understanding the germ of truth that was in them. Later investigators have increased the precision of our knowledge, until we can now say that the present direction of the solar motion is known within very narrow limits. A tiny circle might be drawn on the sky, to which an astronomer might point his hand and say: "Yonder little circle contains the goal toward which the sun and planets are hastening to-day." Even the speed of this motion has been subjected to measurement, and found to be about ten miles per second.

The objective point and the rate of motion thus stated, exact science holds her peace. Here genuine knowledge stops; and we can proceed further only by the aid of that imagination which men of science need to curb at every moment. But let no one think that the sun will ever reach the so-called apex. To do so would mean cosmic motion upon a straight line, while every consideration of celestial mechanics points to motion upon a curve. When shall we turn sufficiently upon that curve to detect its bending? 'Tis a problem we must leave as a rich heritage to later generations that are to follow us. The visionary theorist's notion of a great central sun, controlling our own sun's way in space, must be dismissed as far too daring. But for such a central sun we may substitute a central centre of gravity belonging to a great system of which our sun is but an insignificant member. Then we reach a conception that has lost nothing in the grandeur of its simplicity, and is yet in accord with the probabilities of sober mechanical science. We cease to be a lonely world, and stretch out the bonds of a common relationship to yonder stars within the firmament.

www.ingramcontent.com/pod-product-compliance
Lightning Source LLC
Chambersburg PA
CBHW071424180526
45170CB00001B/208